EXPLAINING CANCER

EXPLAINING CANCER

Finding Order in Disorder

Anya Plutynski

OXFORD
UNIVERSITY PRESS

OXFORD
UNIVERSITY PRESS

Oxford University Press is a department of the University of Oxford. It furthers
the University's objective of excellence in research, scholarship, and education
by publishing worldwide. Oxford is a registered trade mark of Oxford University
Press in the UK and certain other countries.

Published in the United States of America by Oxford University Press
198 Madison Avenue, New York, NY 10016, United States of America.

© Oxford University Press 2018

First issued as an Oxford University Press paperback, 2022

CIP data is on file at the Library of Congress
ISBN 978-0-19-996745-2 (hardback)
ISBN 978-0-19-764250-4 (paperback)

CONTENTS

ACKNOWLEDGMENTS

I am enormously grateful to many people for their support of my work on this book. First and foremost, I wish to thank my family for their support throughout the process of writing the book. I also owe an enormous debt to Dr. Saundra Buys, Dr. Ed Nelson, and the medical staff at Huntsman Cancer Center for making this book possible in more ways than one and to the faculty at University of Utah's Medical School for permitting me to sit in on lectures in 2010, 2011, and 2012. Thanks also to the Medical School faculty at Washington University in St. Louis for permitting me to attend lectures in the Oncology Pathways Series.

I also owe a debt to the many readers who have provided valuable feedback: notably, Joshua Rubin, Carl Craver, Paul Griffiths, Karola Stotz, Jim Tabery, Erika Lorraine Milam, Marc Ereshefsky, Dominic Murphy, Sean Valles, David Teira Serrano, Justin Garson, Ron Mallon, the students in Washington University's fall faculty book seminar (Maria Doulatova, Tomasz Wysocki, Rachel Williams, Shiyu Zhang, Mark Povich, Allert van Westen, Cameron Evans, Michael Carver and Dylan Doherty), and three anonymous reviewers for Oxford University Press. In addition, I owe an enormous debt to the authors of the many books and articles I have drawn upon in my research. This book is a synthesis of the empirical and theoretical work of many scientists, historians, and philosophers of biology and medicine who have had an influence on my thinking: Bissell, Crespi, Weinberg, Merlo, Greaves, Hausman, Waters, Sarkar, Godfrey-Smith, Burian, Solomon, Mitchell, O'Malley, Dupré, Woodward, Sober, Moss, K. Smith, Boyd, Wimsatt, Okasha, Weber, Craver, Cranor, Douglas, Steel, Gannett, Magnus, Morange, Schwartz, Garson, Broadbent, Bechtel, Germain, Bertaloso, and Malaterre. Last but not least, I owe a debt to my editor, Asheligh Imus, Mary Becker, as well as Peter Ohlin and the helpful staff at Oxford University Press.

EXPLAINING CANCER

INTRODUCTION

If you wanted to successfully invade a complex organization, what would you do? You would:

- Start small and grow slowly.
- Begin your work in places weakened by long struggle.
- Be inconspicuous, or, seek to resemble those around you.
- Make use of the physical and communication infrastructure available in your environment.
- Represent yourself as a legitimate member of the organization, when in fact you are using a front to support your own nefarious interests.
- Build up an arsenal of resources for resistance before you're discovered or your opponent engages in any defensive attacks.
- Find ways to expand and diversify your organization, developing specialists in different roles—defensive and offensive.
- Recruit your neighbors or, if they fail to cooperate with you, gradually make it difficult for them to carry on.
- Send out advance parties for invasion, in disguise.

Many cancers[1] fit this description strikingly well. Cancer begins with minor modifications to the body's cells and tissues. It can take decades for a population of incipient cancer cells to develop into an invasive tumor. Tissues long subject to environmental insults, infections, and inflammation are more likely than other tissues to develop a cancer (Grivennikov et al., 2010). Cancer cells co-opt the physical and communication infrastructure in their environment. They send signals akin to those of a healing wound—signals that attract a blood

---

1. Carcinoma is the most common form of cancer. Carcinomas begin in epithelial tissues, such as the colon, skin, and breast. Leukemia, lymphoma, and sarcoma are distinct kinds of cancer that originate in different cell types. While they resemble carcinoma in some respects, the description above of invasion and metastasis is drawn largely from work on carcinomas. Please see Appendix for detail.

supply, induce the secretion of growth factors, and reshape the architecture of surrounding tissues in ways that permit invasion and metastasis (Egeblad, 2010). Tumors are sometimes said to "recruit" surrounding healthy cells and tissue to their cause; these cells acquire various "cancer-associated" properties that enhance a tumor's capacity to grow and, eventually, invade (Hanahan and Coussens, 2012). Cancers also may be said to acquire an arsenal for resistance: as cells divide, they acquire a reservoir of mutations, many of which may play important roles in enabling the cell to block signals from the environment that ordinarily prompt cell death, resist or escape the body's immune response, or, eventually, enable resistance to chemotherapy. Metastatic cells attract platelet "couriers" that function as a "disguise," enabling them to evade detection as they spread throughout the body (Egeblad et al., 2010).

This strategic description makes it enormously tempting to describe cancer as a kind of agent, as if cancer progression were a product of intention or design. Indeed, some argue that cancer is a product of design, but there is no designer, only the same mindless process that yielded the compound eye and tetrapod limbs: natural selection (Merlo et al., 2006; Greaves and Maley, 2012; Greaves, 2001). Insofar as cancers consist of populations of cells with heritable variation, and this variation affects the relative survival and reproductive success of cells and cell lineages, the process of cancer progression described above can be seen as a product of the process of natural selection. Different cells and cell lineages with distinct properties may be more or less successful at growing, and ultimately invading and colonizing metastatic sites. These metastases may in turn be more or less successful at seeding secondary metastases, in part due to heritable variation and in part due to a kind of "niche construction," or the shaping of the environment in ways that, in turn, shape the trajectory of change in the evolutionary process (Barcellos-Hoff et al., 2013).

Many readers may find this picture of cancer counterintuitive, for cancer is typically understood to be due to a breakdown in function. Indeed, cancer is sometimes described as a "failure" in the functional organization of tissues or a "breakdown" in the mechanisms that ordinarily maintain this precise functional organization, akin to the breakdown of a car over time. There is surely something correct in this view, for the incidence of cancer increases (for the most part) with age much as cars break down over time. Our cells divide over our lifetimes, and errors in the replication of DNA during cell division accumulate. These errors contribute, over time, to the risk of cancer developing in given tissues. Tissues with higher rates of turnover of specific types of cells (somatic stem cells) accumulate more mutations and thus, according to Tomasetti and Vogelstein (2015), are subject to higher rates of cancer. This is just what is expected on the "breakdown" picture of cancer. Just as the breakdown of a car is due to the parts' failure to perform their proper functions, cancer cells fail to perform functions typical of normal or healthy breast, prostate, and brain cells.

Which view is correct? Are cancers a process of natural selection, or are they a breakdown in functional organization? This question poses a false contrast. In my view, the strategic and breakdown views of cancer are not at odds. While from the perspective of the organism as a whole, cancer is a failure, from the admittedly short-term perspective of cancer cells, cancer progression may also be viewed as akin to a selective process. Cancer is like a car breaking down, at least in some respects, but significantly unlike a car breaking down in others. The differences are illuminating; they tell us something important about organisms and how they fail, how organisms are like and unlike artifacts, and thus also how biological explanations of disease are like and unlike the explanations offered by car mechanics.

Organisms are distinctive in a variety of ways. First, organisms have a history—they are a product of phylogeny and ontogeny. Cars, while they do have a phylogeny (in a sense), are not a product of ontogeny. Cars do not develop from one or a few cells or change in organization and response to novel environments over time. They do not play an active role in their own development, let alone reshape their environments. While cars in some sense bear "information" about their ancestry, they do not draw upon that information in ways that affect their current organization and behavior, nor do they directly affect the relative proportion of their descendants in future generations. Organisms do, which is why our evolutionary and developmental history helps to predict and explain when, where, and how our bodies break down. This is why understanding our developmental and evolutionary history is essential to understanding disease, and cancer in particular. Current function *and dysfunction* are products of our evolutionary and life history: "Through failure, we understand design" (Frank, 2007).

Second, cars require the ministrations of mechanics, whereas organisms both maintain and (to some extent) fix themselves. Of course, no organism is capable of perfect self-maintenance and repair. Cancer is direct evidence of that (as is any degenerative disease or any disease of aging). Nonetheless, organisms engage in self-maintenance via various feedback mechanisms. Many of these mechanisms are redundant, which is what one expects of evolved systems. If one repair pathway breaks down, often another comes online. It turns out that these evolved features of organisms also explain various facts about cancer. Redundancy and duplication of function play an important role in preventing cancer, but they also explain, in part, cancer's unique ability to evade destruction. Cancer requires the breakdown of a number of typical capacities of the cell and surrounding microenvironment, and there are often "backup" mechanisms that halt the progression to cancer. On the flip side, however, redundancies may also play a role in the ability of cancer cells to survive. For instance, drugs that target one variant of a gene may be incapable of blocking the activity of duplicate genes with similar functions, thus enabling cancers to resist chemotherapy (Yarden et al., 2012).

Third, organisms are hierarchically organized; they have parts within parts, and systems of functional organization within other systems. In this, they are not dissimilar to cars. However, the relationships between the parts of an organism change over the course of development, and indeed, as cancers develop. Parts unique to one subsystem can be redeployed to perform different functions in different systems. This multi-use architecture contributes to various trade-offs in fitness. What may be an adaptive trait at one stage in development or life history (the life course extending from birth through adolescence, middle age, and death) may become maladaptive at another. Traits that are advantageous, and perhaps optimize fitness, at one stage of development may increase the risk of disease later in life (Crespi and Summers, 2005).

Unlike cars, cancers have a distinctive ontogeny—or developmental history. Many cancer cells behave in many ways like cells in earlier developmental stages. For instance, in leukemia there is a kind of "arrest" of blood cells' typical developmental trajectory. Some cancer cells acquire the ability to activate or "access" parts of the genome that ordinarily are "shut down" over the course of development. For instance, for metastasis to be possible, metastatic cells need to "reverse" their developmental specification as epithelial cells (like skin or the lining of the throat, stomach, or colon) to move about without the strictures typical of cells of the epithelium. This is called the "epithelial–mesenchymal transition" and is a reversal of the process typical in development (Weinberg, 2013). This distinctive cellular diversity and plasticity in metazoans explains why plants do not "get cancer" (or, at least, not in the same way as other organisms), because though they develop tumors, the cells are far less plastic and rarely metastasize (Aktipis et al., 2015).

Last but not least, the sex, species, family, or genus to which an organism belongs shapes how, when, and why the organism is more or less vulnerable to disease, and cancer in particular (Aktipis et al., 2015). For instance, there are patterns of association between sex and cancer incidence and mortality, and thus the evolution and development of sex-specific traits may explain in part distinct patterns in cancer etiology (Sun et al., 2015; Crespi and Summers, 2005). Larger body size and longer life span are associated with more somatic cell divisions during an organism's lifetime. It follows that larger animals may be expected to have higher rates of cancer or, perhaps, better ways of preventing cancer. This prediction is borne out in at least one interesting case: whereas humans have only one copy of *TP53* (a gene associated with tumor suppression), elephants have twenty copies of this same gene, which may have evolved to serve a protective function, given elephants' relative size and long life span (Abegglen et al., 2015). Of course, body size and rate of somatic cell division are not likely to have a direct, linear relationship to cancer risk. For tissue architecture and the presence of certain kinds of cells (tissue-specific somatic stem cells) also play an important role in shaping when, where, and how cancer arises in different tissue types, across different species (Laplane, 2016). This may well be an example of another adaptation to the intrinsic risk of cancer for all multicellular organisms (Crespi et al., 2005).

It's easy to forget that the cooperative organization between cells in multicellular organisms is a relatively recent evolutionary innovation. The formation of the first eukaryotic cell and (several million years later) multicellular organisms was an extended process of evolution of cooperative organization. The transition to multicellularity required that cells compromise their individual fitness in service of the collective, though presumably participation in collectives had its advantages (Michod, 2006, 2007). If we look at the tree of life, we can see how cooperation and conflict within collectives are a matter of degree. And we can find instances of more or less cooperative groups in the biological world. Interactions in biological collectives can result in mutualism, cooperative organization, parasitism, or mutual destruction. Collectives with more or less functional organization (such as microbial biofilms, beehives, and humans) and degrees of cooperation among parts might be characterized as having more or less "organismality" (Queller and Strassman, 2009). Cancer may be viewed as an intermediate case, akin to an evolving population of bacteria (Gauss, 1966; Godfrey-Smith, 2009a).

In sum, there are several ways in which our vulnerability to disease is shaped by our evolutionary and developmental history, and thus there are several reasons why explaining disease is not quite like explaining the breakdown of a car. First, organisms, unlike artifacts, have an ontogeny and phylogeny, and these historical processes in part shape how and when we fail. We have various features and capacities (modularity, redundancy, plasticity) that enable self-maintenance. We are subject to fitness trade-offs, and our vulnerability to cancer is in part a byproduct of such trade-offs. More generally, evolutionary history can assist in predicting and explaining species' relative vulnerability to disease, and cancer in particular. In sum, our health is in many ways a delicate compromise of interests, a balance of competing goals, constantly undergoing intrinsic and extrinsic changes in stability, composition, and organization over time.

There are three upshots of this discussion. First, it often happens in science that two ways of thinking about a class of phenomena that may prima facie seem at odds can be shown to be consistent, provided we narrow or make more precise their domain of application. Second, theoretical perspectives or models are always (inevitably) partial in their explanatory scope. Third, cancer is such a heterogeneous kludge of many different diseases, and is so causally complex, that it may be examined from different angles and can be analogized to almost anything.[2] Cancer may be viewed as akin to a car breaking down, a process of natural selection, an infectious disease (Liu et al., 2016), a process of development, and even the growth of an ecological community.

Scientists often deploy metaphors and analogies like these as tools for investigation, to inspire novel hypotheses (Hesse, 1966; Campbell, 1920). Some have claimed that the price of deploying metaphor is eternal vigilance.[3] However, scientists do not intend analogical models to represent the system of interest in all respects, or even entirely accurately in the respects intended. "As if" thinking enables scientists to formulate hypotheses, make predictions, and increase their understanding. What precise relation ought to hold between a model and the system of interest depends on what scientists intend to use the model *for* (Morgan and Morrison, 1999; Teller, 2001; Giere, 1999). Imperfect knowledge of the respects in which or the degree to which our model fits the world is to be expected. Indeed, mapping the contours of what we don't know may be exactly what we are interested in pursuing with "as if" thinking.

No model can possibly represent the world in all possible respects. Rosenblueth and Wiener made this point rather well almost seventy-five years ago:

> Let the model approach asymptotically the complexity of the original situation. It will tend to become identical with that original system. As a limit it will become that system itself . . . Any one capable of elaborating and comprehending such a model in its entirety would find the model unnecessary, because he could then grasp the universe directly as a whole . . .
>
> This ideal theoretical model cannot probably be achieved. *Partial models, imperfect as they may be, are the only means developed by science for understanding the universe.* This statement does not imply an attitude of defeatism but the recognition that the main tool of science is the human mind and that the human mind is finite. (1945, 320–321, emphasis added)

Rosenblueth and Weiner are resisting the extreme of what van Fraassen called "naïve realism": the view that "the picture science gives us of the world is a true one, faithful in its details," or that the aim of science is to recapitulate the world itself, in exacting detail. Even our best theories may aspire to capture the truth

---

2. I will say more about this later, but roughly, this is the idea that we can decompose cancer in a variety of ways and identify causal processes and relationships at a variety of temporal and spatial scales. For discussions of causal complexity, see, e.g., Wimsatt (1972) and Mitchell (2003, 2009); more recently, Bertolaso (2016) argues that cancer is causally complex as well.

3. Lewontin (2001) attributes this claim to Norbert Weiner, in a paper coauthored with Rosenbleuth in 1945, but no such claim appears in the paper.

only in some respects or to some degree; indeed, sometimes building a false model is the best way to get to truth (Wimsatt, 1987). Model-building often involves deliberate falsehood or idealization. Phenomena of interest may be represented as a product of a set of variables in a directed graph, as a mathematical equation, or as a lattice or network; each choice may serve a distinct purpose or several purposes.

Perhaps especially for complex phenomena such as cancer, it seems that scientists out of necessity deploy a variety of partial representations, each of assistance in addressing only one set of questions about the phenomena of interest. A model that captures important causes of cancer at one temporal and spatial scale will often black box (or, leave for later elaboration) causal variables at another. For instance, the past twenty-five years of research in the cell and molecular biology of cancer has been enormously successful in achieving its central aim: identifying specific mechanisms and pathways associated with the distinctive behaviors of cancer cells. Disrupting these pathways has been shown to lead to the "hallmarks" of cancer—the capacity for limitless, unregulated growth; evasion of cell death; and the capacity to acquire a blood supply, invade, and metastasize (Hanahan and Weinberg, 2000). However, this approach to cancer tends to focus attention on one relatively narrow temporal and spatial scale. One of the central aims of this book is to argue that understanding cancer requires both the decomposition of parts and processes involved in cancer at the cell and molecular levels ("drilling down") and "scaling up" to the macro level, or examining cancer's historical origins and remote causes, complex organization, and dynamics. That said, different methods may be fruitfully deployed in service of distinct aims. The best tool for the job depends to an important extent on what we wish to predict or explain or how we wish to intervene, as will become clearer over the course of discussion in the chapters to come.

1.1. Organization of the Book

Chapter 1 begins with the question of whether cancer is one kind or many. If many, then how many, exactly? I argue that there is no single "natural" classificatory scheme for cancer. Rather, there are many equally natural ways to classify cancers, which serve different predictive and explanatory goals. Cancer classifications are investigative categories, which serve a variety of scientific and practical purposes (see Brigandt, 2002; Griffiths, 2004). To set the stage for my argument, I consider two philosophers' views concerning whether cancer is a natural kind. First, Khalidi (2013) argues that cancer is the "closest" any scientific kind comes to a homeostatic property cluster (or HPC) kind. While it is prima facie persuasive, the HPC account suffers from both regress and ambiguity problems. For on the HPC view, we first need to find a principled way to pick out or demarcate "mechanisms"; and second, we need to determine what suffices for a mechanism to count as "homeostatic." I argue that even were we to agree upon "the" mechanisms for cancer (which, in my view, is a pragmatic matter at best; cf. Craver, 2009), whether these mechanisms are homeostatic is far from obvious. However, viewing cancer as a single disease with a common underlying mechanistic basis—or at minimum, shared features—was nonetheless a fruitful research program. Lange (2007) arrives at the opposite conclusion from Khalidi; he argues that cancer is at best a "kludge," and advances in molecular subtyping of cancer hail the "end of diseases" as natural kinds. I argue that while molecular and biochemical features of a tumor and patient are useful and important to know, they are not sufficient to serve scientists' and clinicians' diverse explanatory and predictive purposes. I then use this case study to consider the more general question of what it means for a classificatory category to be "natural" and what it means for it to be "scientific." I consider several alternative accounts of natural or scientific kinds, the "simple causal view," the "stable property cluster" view, and "scientific kinds" (Craver, 2009; Khalidi, 2013; Slater, 2014; Ereshefsky and Reydon, 2016). These better accommodate diverse modes of classification deployed in scientific practice. However, they also illustrate how the line between natural and non-natural kinds may be impossible to draw in a domain-general way (cf. Slater, 2014). Scientists' goals and thus criteria for classification are often far more domain-specific than philosophers have assumed; the general category of "natural" kinds is not, after all, terribly natural (cf. Khalidi, 2013; Ereshefsky and Reydon, 2016).

In Chapter 2, I consider several "line-drawing" problems. First, in the diagnosis of early-stage cancer, there is an epistemic or "inductive" risk (Douglas, 2000) involved in assessing whether a suspicious lesion is likely to progress to metastasis. Cancers have variable courses of progression; so the assessment of disease state as a practical matter is not strictly an empirical judgment, but also an evaluative one. In borderline or contentious cases, tumor boards (groups of clinicians, including oncologists, pathologists, and surgeons) together make recommendations for patients about the appropriate level of precaution in conditions of uncertainty. This raises some interesting questions; in particular, in such ambiguous cases, what role ought patients' values play in treatment decisions, and what really constitutes "shared decision-making" in medicine? Second, when, whether, and how often to screen is another case where the evidence underdetermines judgments about "effectiveness." I argue that there are many sources of evidential underdetermination in the epidemiology of cancer screening. So here as well, we have a case of inductive risk and a line-drawing problem. Last, and more generally, these cases of early-stage cancer raise interesting questions about the line between health and disease, as it often involves assessing the risk of future harm, or what Aronowitz (2015) has called "risky medicine." Naturalistic accounts of disease underdetermine our choices of how or where to draw the line between health and disease, in such cases. Indeed, some naturalists have incorporated into their account the fact that compromises to health or "healthy functioning" are a matter of degree (see, e.g., Schwartz, 2007, 2008; Hausman, 2014). I argue for a mixed view that incorporates both empirical and normative judgment and risk assessments; this view, I claim, makes more transparent when and how values enter into judgments about disease.

Genomic medicine has lately been hailed as a tool for cancer prognosis and treatment, potentially resolving some of the uncertainty about which cancers are more likely to respond to which treatment, and perhaps also which early-stage cancers are likely to be more (or less) aggressive. Differences in gene expression among cancers will, on this view, be precise predictors of prognosis and treatment response. While this approach has great promise, in Chapter 3 I confront a variety of questions that the approach raises as to how we define and explain cancer. Drawing upon useful distinctions some philosophers of science have drawn between causal relevance, strength, and specificity (cf. Hausman, 2010; Woodward, 2010), I argue that while genes are causally specific, they are by no means the only causally specific factors in cancer etiology. As a backdrop to this discussion, I introduce what I call the "mechanistic research program" in cancer, according to which progression to cancer involves breakdowns in regulatory controls on gene expression in ways that affect cell birth and death. Disorderly cellular growth is affected, however, by many pathways, and mutations to genes like *src, ras, myc, APC, p53*, and others are only one of many nodes in these pathways. Mutations to "cancer" genes play *a* causal role in the progression to cancer; but the causal role these mutations play is strongly context-dependent. That is, mechanisms for cancer are unstable, highly context-dependent, and local in their effects. Many causes at and above the cellular level act and interact in cancer, and so explaining cancer requires a more integrative, multiscale approach. I place this discussion in the context of what philosophers have called the "causal selection" problem. I conclude by considering several accounts of causal selection and argue that ultimately the causal selection problem is not one but several different problems, requiring different and context-specific solutions.

In Chapter 4, I consider two case studies in environmental epidemiology of cancer, in service of three aims. First, I argue that while mechanisms are useful to have in epidemiology, they are neither necessary, nor sufficient (it turns out), to justify claims about causal regularities at the population level where mechanistic bases are multiply realized. Second, I argue that Hill's famous "criteria" (1965) for determining whether epidemiologists are justified in inferring an environmental cause of cancer can best be considered a regulative ideal: not "criteria" for judging that X is a cause, but rather an argument for consideration of total evidence in conditions of uncertainty. Third, I argue that standards of evidence for establishing causation ought in part to hinge upon what we intend to use the evidence for. Since epidemiological data is so important for the regulation of carcinogens, the practical matter of what the causal information is *for* must be taken into account in setting standards of assessments of evidence. Cranor's (1997) discussion of risk assessment shows that if we care more about social welfare than accuracy, we ought to choose the method that is more efficient at getting results quickly. Indirect but still strongly suggestive evidence of a causal link may be good (enough) reason to take precautionary measures, independently of our empirical judgment that such evidence is decisively in favor of a causal link. That values are taken into account here does not, in my view, compromise objectivity. Indeed, making clear when and how values play a role is part of what is required to make both the science and law and policy more objective and transparent. My argument draws upon a variety of philosophers' views on precaution, risk assessment, and the role of values in science and policy (Longino, 1990; Mayo, 1988; Douglas, 2009; Mitchell, 2009; Steel, 2014; Cranor, 2011; Reiss, 2015). This argument dovetails nicely with Reiss's (2015) "pragmatic" account of evidence. While Mayo's error-statistical approach may in some cases provide a quantitative measure of strength of evidence for environmental cancer risk, unfortunately these are only ideal cases, where we can do certain kinds of experiment. In less ideal cases, we must appeal to a variety of independent sources of evidence, along the lines Hill suggests. In such cases, there is no straightforward way of measuring strength of confirmation; however, I argue that independence of evidence should give us greater confidence in a hypothesis.

In Chapter 5, my goals are threefold. First, I give a taxonomy of evolutionary explanations of cancer, illustrating distinct types with case studies of explanations of cancer's origins and prevalence. Second, I defend a multilevel perspective, arguing that cancer progression is a process of selection and a byproduct of selection at distinct levels in the biological hierarchy. Third, I provide a general account of the explanatory role of evolutionary dynamic models of cancer progression and treatment failure. I use these accounts to illustrate examples of "how possibly" and "a priori" causal modeling, as defended by Sober (2011), and respond to critiques from Lange and Rosenberg (2011).

In Chapter 6, I draw upon several twentieth-century historical case studies to argue that cancer research is not theory-driven but problem-driven. That is, cancer researchers have been largely concerned with puzzles, some more general and some more specific, a concept I draw in part from Kuhn (1962). Examples of several puzzles serve as illustrations: the puzzle that cancer on average increases with age, the puzzle of why we don't get cancer more often than we do. I argue that what are sometimes called "theories" of cancer are best viewed as ways of framing (more or less fruitful) "research programs" (Lakatos, 1980). Research programs are successful to the extent that they set out problems or puzzles worth solving.

In sum, I argue that because cancer is so descriptively and interactively complex (Wimsatt, 1972), "explaining cancer" requires the viewpoint of different disciplines and different kinds of representations. This chapter thus defends a pluralist and pragmatic approach to scientific explanation in the biomedical sciences. Explaining in science is an iterated and ongoing process, and understanding is a matter of degree. Models (and explanations) of cancer are often partial and sometimes overlapping, as has been argued in other contexts by Mitchell (2003) and Massimi (2016).

I.2. Autobiographical Note

I began this book about eight years ago, when I was diagnosed with breast cancer. I was deeply puzzled by my diagnosis, for several reasons: I had no family history of cancer, or breast cancer in particular. I was not a smoker, nor had I been exposed for any length of time to various risk factors for cancer (as far as I knew). I was a healthy woman, in my mid-thirties. I had benefited from plenty of opportunity to live and work in healthy environments, and was, as far as I knew, free of risk factors for cancer. So why did I get cancer? Was this simply a matter of bad luck? Had I done something that led to this cancer diagnosis? What role—if any—did my actions and choices play?

This question led me to yet further questions, which led ultimately to this book. I spent several years taking courses at the medical school at the University of Utah and at Washington University in St. Louis. I am immensely grateful to the scientists and clinicians who had the patience to answer my questions and permitted me to sit in on their lectures, lab meetings, and seminars. The book is (roughly) structured around the (initially naive) questions I had about cancer, growing out of my personal experience: What is cancer? How is cancer distinctive as a disease? How do scientists investigate cancers' causes? When and why do they claim to have good evidence for this or that cause of this outcome? Is there a "general theory" of cancer causation? Can we even "explain cancer" in general? Or is the search for general explanations of cancer misguided?

Of course, there are no simple answers to such questions. Thousands of papers are published annually on cancer, and millions of dollars are spent annually to investigate cancer from a variety of scientific perspectives. No one book could or should attempt to summarize all this important work. What I have done instead is to tie my initially naive questions about cancer to broader questions in philosophy of science, in service of starting a conversation among philosophers of science, scientists, and science studies scholars, more generally. Namely:

- How do scientists classify cancer? Do these classifications reflect nature's "joints" or something else? If something else, then what, exactly? Ought we to expect the variety of modes and methods of classification that cancer scientists and pathologists use to become unified, or derived from a common "essential" set of properties or features of cancer types and subtypes? Why or why not? What does this case teach us about natural classifications in science?

- Is there a sharp line between disease and "indolent disease" or "pre-disease"? In borderline cases, where the evidence underdetermines pathologists' judgments, how—if at all—do values play a role in such judgments? What can the case of early-stage cancer teach us about debates in medicine about the nature of disease and how we ought to demarcate disease and health? What is the difference between disease and disease "risk"? What can borderline cases of cancer teach us about the role of values in science and medicine?

- What does it mean to say that cancer is a "genetic" disease? Why do cancer scientists pick out genetic mutations as such important causal factors for cancer? Are there "genes for" cancer?

- What are the most important environmental causes of cancer, and how do epidemiologists investigate these causes? What can environmental epidemiology teach us about evidential reasoning in science in general? How are the adequacy conditions on evidence tied to (or how ought they be tied to) what we wish to use our evidence for?

- How exactly has our evolutionary history made us especially vulnerable to cancer? Is cancer itself an evolutionary process or a byproduct or both? What can modeling cancer from an evolutionary perspective teach us about modeling in science more generally? Is there one "best" theory of cancer, or only many different models?

Standing behind all these questions is a larger question: What does it mean to "explain" cancer? One's first response to such a question might be to point out that the question is vague. There are societal, economic, institutional, environmental, and of course biological causes of cancer. Which of these many causes provides "the answer" to this question? Perhaps there is no single general explanation for "cancer" broadly understood. A second response is the following: this is an empirical question. What role, if any, might philosophers have in service of addressing such a question?

Many scientists may be familiar with philosophers of science like Karl Popper and Thomas Kuhn. They may have followed various public debates about teaching evolution in public schools, and so may be familiar with philosophers of science who address questions that such cases raise about how to demarcate science from pseudoscience. There are some pat answers to these questions—for instance, Popper argued that falsifiable hypotheses and crucial experiments are essential to scientific inquiry. However, as the case of evolutionary biology makes rather vivid, many of the questions scientists investigate cannot be resolved by crucial experiments.

That does not make them unscientific. No crucial experiment can establish that all life descended from common ancestry, but a wide variety of independent evidence does support the fact of common descent. Evolutionary biologists do conduct experiments, but not all our understanding of the history of life on earth is derived from experimental methods. Most of the history of life occurred millions of years ago, and so what and how we come to know about it require very different scientific methods—the discovery and investigation of fossils, comparative biology, mathematical modeling, computer simulation, and the fitting of both historical and current data to various predictions of our models. In other words, this is a case where Popper's model of science doesn't quite capture the nature of inquiry. Rather, inference to the best explanation, drawing upon the widest possible array of independent evidence, is the "inference that makes science" (McMullin, 1992).

Like life on earth, cancers often bear only traces of their origins. Thus, cancer scientists often have to make indirect inferences about the history of this disease from investigations of properties of tumors or metastatic disease in deceased patients, investigations in cell culture, or experiments with model organisms. Cancer cells in culture and model organisms are only imperfect models of cancer in vivo. This is only part of the reason why the investigation of cancer, in general, is a hard problem. There are also a variety of causal factors of relevance to cancer, operating on a variety of temporal and spatial scales—from biochemical interactions at or below the level of the cell to the evolutionary timescale. So cancer scientists must draw upon diverse disciplines to draw inferences about how cancer is caused and might be prevented. Fitting all these perspectives together can be a challenge, to say the least. The complexity of cancer, and of cancer science, requires that we rethink the picture of science inherited from philosophers like Popper.

What picture of science ought we to adopt? We tend to think of scientists as aiming at general "unified" theories; but cancer, and cancer science, forces us to rethink this picture. For instance, I have said that cancer is "complex." How exactly is cancer complex? There are at least two senses of complexity at issue here: descriptive complexity and interactive complexity (Wimsatt, 1972). To unpack the idea of descriptive complexity, we need to first understand the idea that an entity, a system, or a process may be "multiply decomposable" (Wimsatt, 1972). A multiply decomposable entity, process, or system is one that can be broken down into parts and investigated in different ways (Simon 1962; Kauffman, 1971):

> [A] system can be viewed from a number of different perspectives, and that these perspectives may severally yield different non-isomorphic decompositions of the system into parts . . . systems for which these different perspectives yield decompositions of the system into parts whose boundaries are not spatially coincident are properly regarded as more

descriptively complex than systems whose decompositions under a set of perspectives are spatially coincident. (Wimsatt, 1972, 69)

Wimsatt produces a vivid illustration of two cases, one of which is descriptively simple and one complex.

As we can see from Figure I.1, it is possible to decompose and investigate the fruit fly in at least six different ways—ways that focus on different aspects or features: its physico-chemical composition, its anatomy, its cell types, its development, the rate and type of biochemical reactions involved in this system, and the network of relationships between physiological systems. Each such investigation requires distinct "decompositions" or demarcation of the relevant parts of the fly and of their causal and spatial relations. In contrast, although we could also investigate, for instance, the chemical composition of a rock, it's thermal conductivity, electrical conductivity, density, and tensile strength, all of the appropriate or relevant decompositions of the rock in service of these questions result in more or less isomorphic ways of "breaking down" the rock into relevant parts (or so Wimsatt claims—arguably this is contestable; see, e.g., Bursten, 2016).

In contrast, "interactional complexity" concerns how these ways of breaking down a system into parts are interrelated. Herbert Simon (1962) argued that "nearly decomposable" systems were hierarchical systems where interactions between subsystems were weak but interactions within each subsystem were strong. More precisely, nearly decomposable systems are those where "the short-run behavior of each of the component subsystems is approximately independent of the short-run behavior of the other components," and "in the long run, the behavior of any one of the components depends *in only an aggregate way* on the behavior of the other components" (Simon, 1962, 474; emphasis added). Simon argued that we should expect complex systems to have a hierarchical structure and to be nearly decomposable. Wimsatt disagrees:

> Simon's use of the concept of near-decomposability . . . sometimes appears to suggest he believes such hierarchical systems to be nearly-decomposable in a nestable manner—with smaller subassemblies (at lower levels) having successively stronger interactions . . . To accept this opinion is to fail to distinguish the decomposability or stability of the subassemblies before they aggregate from their decomposability or stability (in isolation) after they have aggregated . . . when they have had time to undergo a process of mutually coadaptive changes under the optimizing forces of natural selection . . .
>
> [T]he optimizing effects of selection [create more interdependence, specialization of function, and interdependence of parts] . . . so that hierarchically aggregating systems will tend to lose their neat S-decomposability by levels, and become interactionally complex. (1972, 77)

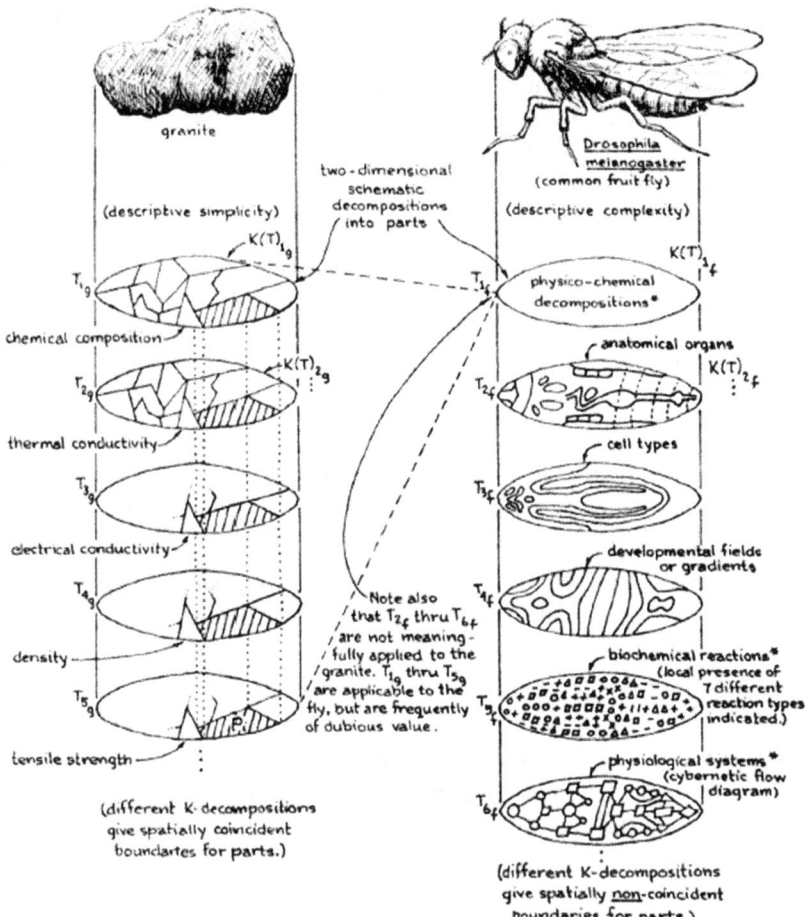

FIGURE I.I Wimsatt (1972) distinguished "descriptive" and "interactional" complexity. Descriptively simple systems can be decomposed into parts in ways that are spatially coincident. Descriptively complex systems have non-isomorphic, non-spatially coincident decompositions. This image of two decompositions illustrates a descriptively simple system (granite), and a descriptively complex one (*Drosophila*). In this chapter I argue that cancer is an instance of the latter.

This passage is difficult to interpret but is worth unpacking, as Wimsatt's point here is central to one of the arguments of this book. Simon thought about biological systems somewhat ahistorically; he was thinking of them as finished systems, like computers, that were hierarchically organized, with relatively discrete parts and subparts, whose behavior can be more or less predicted by aggregating the effects of each part and subpart. For artifacts like computers or clocks, Simon's assumptions about the short-run independence and merely aggregate dependence of subassemblies (or near decomposability) may well be correct. However, when evolution is acting on a complex system, the neat hierarchical picture that Simon gives generally breaks down, according to Wimsatt. Evolution can lead to what Wimsatt calls "interactional complexity." This is where interactions between the components of different subsystems affect their individual and collective behavior over the short term; and the long-term properties and dynamics of a system are not simply a product of the aggregate activity of subassemblies.

Wimsatt argued, contra Simon, that evolved systems tend to become interactionally complex, and we should take this seriously when we attempt to explain and describe them. For instance, it's typical in biology to decompose an organism into parts that play discrete functional roles—metabolism, immune response to infection, and the like. But biological systems are not built by computer programmers, but by evolution, and evolution is a "tinkerer" (Jacob, 1977). By this, Jacob meant to emphasize that evolution recycles; the same parts that may be proper parts of one subassembly, in service of one function, may be redeployed in service of different functions or different kinds of outcomes or processes. Rather than starting from scratch, bits and pieces are reused or activated in novel ways in different subsystems, in service of new functions. An organism can be decomposed in different ways, according to different theoretical perspectives or investigations into different features of a system. Interactional complexity is a measure of how much interaction we can find between features of such subsystems. Wimsatt argued that how interactionally complex a system is affects how many theoretical perspectives one must consider for their predictions about the behavior of that system to be accurate.

How does all this bear on cancer and on explaining and understanding cancer? Steve Frank once said (personal communication) that for a cancer scientist with interests in metabolic features of cancer, everything of interest in cancer can be explained by metabolism, whereas for a cancer scientist interested in stem cells, everything of interest in cancer can be explained by stem cells. Frank was making a joke, but it is a telling one; each scientist investigating one of several ways of decomposing the causal factors of relevance to cancer is likely to see such factors as centrally important to many, if not all, aspects of cancer initiation and progression. But, of course, no one scientist is going to give us the whole picture. This could be predicted for descriptively and interactionally complex systems (Wimsatt, 1972). This interactive complexity of organisms has massive implications both for our study of living things and for our study of how they break down.

The appropriate decomposition of parts and processes in service of understanding one particular feature or aspect of an organism's function may overlap with investigations into yet another feature. These overlaps in functional roles of parts contribute (in part) to vulnerability to diseases like cancer. Biological systems can be decomposed in different ways, but these different decompositions have overlapping parts; so understanding how, for example, metabolism affects the growth and invasion of a tumor is *not* entirely independent of understanding how and why tissue organization likewise affects tumor growth and invasion. This interactive complexity means that cancer scientists *must* think of cancer as not only a "genetic disease" but also a disease of tissue organization, a metabolic disease, an environmental disease, a developmental regression, and an evolutionary process. All of these ways of thinking about cancer are not only informative but also mutually informative. Cancer is a vivid instance of a descriptively and interactively complex causal process. Thinking about cancer as an instance of an interactively complex system can help us think more carefully about how to do science, as well as the nature of the biological world and ourselves as part of that world.

CANCER

NATURAL, MEDICAL, AND SOCIAL KIND

Although all physicians know what they mean when they use the term
neoplasm, it has been surprisingly difficult to develop an accurate definition.
(Stricker and Kumar, 2010, 280)
Science has room for both lumpers and splitters. (Sober, 1999, 551)

1.1. Introduction

What, if anything, unifies cancer as a topic of inquiry? Is it merely
a historical accident that all the things we call "cancer" are labeled
similarly? Do all cancers have some features in common? Are these
features sufficient to characterize cancer as a "natural kind"? Is cancer
one kind or many? If many, how many, exactly?

On the one hand, it is surely true that leukemia, lymphoma, and
solid tumors of the lung, breast, and prostate are very different diseases
with different etiologies, prognoses, and so on. One plausible way of
reading the history of how the term "cancer" came to be associated
with such different diseases is as a lesson in contingency or "historical
accident" (Mukherjee, 2010). On the other hand, there are common
pathways affected, or similar types of aberration, across a variety of
cancers: all cancers involve in some way or another aberrant growth or
maturation of cells, distinct from what is typical for a particular tissue
or cell type, whether skin, bone, nerve, or blood cell. This picture seems
relatively unified; there are alterations to common regulatory pathways
that result in typical "capacities" or "hallmarks" of cancer (Hanahan and
Weinberg, 2000, 2011): self-sufficiency in growth signals, insensitivity
to growth-inhibitory (antigrowth) signals, evasion of programmed cell
death (apoptosis), limitless replicative potential, sustained angiogenesis, tissue invasion and metastasis, deregulation of cellular energetics,
genome instability, evasion of immune response, and inflammation.[1]

---

1. For a critique of this picture of cancer, however, see Sonnenschein and Soto (2013).

Yet again, the more we learn about cancer, the more diverse it appears to be. It is becoming increasingly common to hear talk of the genetics and molecular "signatures" unique to each cancer type and subtype. Perhaps then cancer is not one but hundreds of diseases. In fact, some suggest that each cancer is in some sense a unique disease process. It seems that one might make a plausible case for a range of views on the matter of whether cancer is one or many. Which view is correct?

My aim in this chapter is to challenge the presupposition behind this question. It is a mistake to think that there is one correct answer to the question "How many kinds of cancer are there?" There are a variety of equally satisfactory (or, if you like, "natural")[2] ways of classifying and, indeed, cross-classifying cancers. This pluralism of legitimate classificatory schemes is not a product of incomplete science, nor is it merely a reflection of our diverse interests and purposes. That is, I endorse a modest form of realism and pluralism about cancer kinds. Cancer kinds are (defeasible) investigative categories that have proved fruitful for the purposes of prediction, explanation, and intervention. Some assume that if such categories are interest-relative—or depend on scientists' contingent goals and purposes—then they cannot track "real" distinctions. I do not see the fact that scientists' interests shape this enterprise as in tension with modest realism. Indeed, I don't see any alternative for a naturalistically informed, realist philosophy of science.

Cancers have a variety of properties (location, organ, cell, tissue of origin, character of tissue microenvironment, extent of profusion, shared mutations, extent

---

2. This term, needless to say, carries a great deal of baggage. I will address this matter of "naturalness" in the final section of the chapter. Lewis (1983) argued that properties are "more or less" natural. Perfectly natural properties, in his view, are mutually independent and, moreover, constitute a "minimal supervenience base" for everything. What this means is vague; but (roughly) the "natural" properties are those that are independent of all other properties—they are the "basic" features or properties of the physical world. While Lewis is talking of properties, not kinds, many defenders of essentialism, and of microstructuralism about kinds, share with Lewis the idea that scientific kinds are genuinely "natural" only when they are mutually independent, or "fundamental" in some sense, though it's not always made clear which sense, exactly. Putnam (1973) adds to this view the idea that kinds figure in universal and necessary laws. This picture of nature and of what makes things "natural" should seem odd to anyone who knows something about genetics, evolutionary biology, or ecology. One problem (of several) with this view is that human beings are part of the natural world and make a causal contribution to the regularities to be found in nature, which include regularities in which we (human beings) figure (see, e.g., Hacking, 1999). These regularities lack "necessity" and are (of course) not universal. Another is the presupposition that the natural properties are intrinsic and "mutually independent." This is to rule out by fiat most properties in the biological world (see, e.g., Okasha, 2002). In the post-Darwinian age, we should not be surprised to find that the ways in which and reasons why entities in the world share the features they do have to do not only with microstructure, but with facts about location, timing, and circumstance. How a cell behaves, for instance, has to do with not only the "intrinsic" features of a cell, but also where and when (in development, or life history) it finds itself. Likewise for organisms, species, and higher clades.

of intratumoral genetic heterogeneity, shared signaling networks affected, etc.) and associated dispositions. Such properties cluster in ways that allow us to predict how cancer types and subtypes are likely to behave. Cancer researchers' classificatory schemes identify similarities and differences across cancers that (more or less) reliably predict various outcomes: from likely disease course to response to targeted drugs. That said, generalizations about cancer types' and subtypes' typical behaviors are not exceptionless. Moreover, cancer types and subtypes have the properties they do in virtue of a heterogeneous array of causal and structural features. Much like species, cancers are genetically and phenotypically heterogeneous, with distinct ontogenies and phylogenies. And the same cancer may belong to more than one kind, where these kinds do not stand in a nested hierarchical relationship.[3] Nonetheless, these classificatory schemes meet many of the predictive and explanatory ends they are designed to serve. Thus, in the context of cancer nosology, a pluralist realism may seem the most sensible stance.[4]

Defending this view will require defending an account of natural classification that is relatively permissive and an argument that taxonomic pluralism is not in tension with modest realism. In service of these ends, I begin with a succinct overview of the variety of ways cancer is classified in practice. Next is a discussion of the ways in which competing accounts of diseases as natural kinds face similar underdetermination problems. In particular, Khalidi (2013) argues, for instance, that cancer (in general) comes the closest to a homeostatic property cluster kind of any of the kinds he discusses. I argue that "mechanisms for" cancer are pragmatically demarcated; and whether such mechanisms are meaningfully described as "homeostatic" is ambiguous, at best. Lange (2007) argues for a function-analytic account of disease kinds, where diseases are distinctive kinds of disruptions of function. Moreover, he contends that the rise of molecular medicine hails the "end of diseases" (as natural kinds). I contend that there are simply too many functions disrupted in cancer (or cancer types or subtypes) for one classification scheme to pick out the "correct" or "natural" classification. That is, even if we adopt Lange's view, the question as to how many cancers there are, or how best

---

3. Khalidi (1998) calls these "cross-cutting" classifications and argues that they appear to be very common in science.

4. These terms are contested, but, roughly, "taxonomic realism" is the view that our best scientific theories do provide knowledge of a mind-independent reality, in this case about taxonomy. "Conventionalism" is the view that our description is chosen on the basis of a "pragmatic assessment of what best serves our ends in describing the phenomena *and nothing more*" (Chakravartty, 2011). In contrast, "eliminativism" is the view that our classifications ought to be eliminated from our scientific ontology. I will defend a modest form of taxonomic realism here. See also Khalidi (2013); Chakravartty (2007).

to classify cancers, is empirically underdetermined. Moreover, like mechanisms, functions are pragmatically demarcated.

Craver's (2009) "simple causal view" (that "natural kinds are the kinds appearing in generalizations that correctly describe the causal structure of the world") has promise, but to the same extent that our pragmatic interests enable us to pick out mechanisms and functions, they determine which causal generalizations we pick out as important or interesting. I agree with Craver that to the extent that these interests track regularities in the natural world, they identify natural classifications of cancer kinds. But given our interests in diverse outcomes at a variety of temporal and spatial scales, equally "natural" classifications of cancers can be more or less fine-grained. There is nothing more to say about how we ought to "ground" natural kinds—at least in cancer science. The same repeatable features may be "grounded" in very different ways. Let us now turn to the evidence for this claim.

1.2. Cancer Classification in Practice

Consider some examples of generalizations about different cancer types:

- Grade IV glioblastomas are highly malignant and have a poor prognosis.
- The TNM system, based on the depth of tumor wall invasion, lymph node involvement, and distant metastases, roughly distinguishes groups of colorectal cancer (CRC) patients with remarkably different prognoses. The five-year overall survival for stage I is 93%, decreasing to 80% in stage II and 60% in stage III cases.
- On the basis of their expression level, breast tumors can be classified as hormone-receptor-positive, HER2-positive, and triple-negative (ER-, PR-, and HER2-negative). This classification not only is prognostic, but predicts treatment response.

These generalizations describe various types or kinds of cancer and their typical behavior. For instance, grade IV glioblastomas are cancers of the brain that are relatively advanced and are likely to be recalcitrant to even the most aggressive treatment regimen. Stage I versus stage II colorectal cancers are cancers of the colon and rectum that have specific size, extent of invasiveness, and number of metastases, likewise indicating extent of advancement. Breast cancers may be made up of cells that grow in response to the presence of hormones, or fail to do so. Estrogen-receptor-positive (ER-positive) breast cancers grow in response to estrogen. These distinctive features are associated with different prognoses and differential responses to drugs designed to block the activity of hormones. Each of these descriptions of cancer types and subtypes refer to different features of cancer that predict and explain typical outcomes—from risk of death to response to treatment. Such features include location, stage of advancement, usually based on imaging and tumor biopsies, and "biomarkers" (roughly, any and all features of a tumor other than gross histopathology, e.g., proteins expressed, antigens bound, and/or genomic features), based on samples of tumor tissue taken from a patient.

Cancer classification often begins with a general distinction between solid cancers (e.g., carcinomas, sarcomas) and hematologic cancers, or cancers of the blood and lymph systems. Solid cancers are further classified as either carcinomas, which originate in epithelial tissues (the lining of the lungs, colon, skin, or esophagus), or sarcomas, which originate in non-epithelial tissues (such as bone, nervous, fat, or connective tissue). Solid cancers are also classified in light of locale (head and neck cancers, left or right side of the colon), functional system affected (endocrine, gynecologic, urologic, etc.), and specific organ, tissue, and cell of origin (e.g., glioblastomas are brain cancers that originate in glial cells). Within each organ system, there are distinct functional and anatomical structures, containing different tissue and cell structures, from which distinct cancer types and subtypes arise. For instance, there are three basic types of gynecological cancer: cervical, uterine, and ovarian, and ovarian cancers can be serous or mucinous, whereas uterine cancers can be uterine sarcomas or endometrial. Endometrial cancers can be endometrioid or clear-cell. Complicating this, however, is the fact that there are "mixed" types, or tumors that appear to contain distinct cells of origin. These cancers fall outside of typical classificatory schemes and are often rather difficult to diagnose and treat (Scolyer et al., 2010).

Cancer is a process. Thus, as we have seen, pathologists also classify solid cancers by stage and grade. The TNM stages I, II, III, and IV have to do with primary tumor size, extent of regional lymph node involvement, and number of distant metastases. In contrast, "grade" refers to the degree of differentiation of cells, often assessed via the extent of mitotic activity. For instance, in breast cancer, grade is typically assessed via characteristics of cells taken from a tumor biopsy and viewed under a microscope (or, increasingly, assessed via automated machines). Such characteristics might include tubule formation (as an expression of degree of differentiation), nuclear pleomorphism (variations in the size, shape, and staining of the nucleus of a cell), and mitotic counts (how many cells are dividing). Less differentiated cells appear to have lost more of the distinctive features typical of cells of origin, and this indicates a more advanced cancer.

Though grade and stage are assessed differently for different cancers arising in different organs, they represent similar stages of advancement of the disease and, not coincidentally, have similar implications for disease-free survival. These categories are relatively predictively successful, in other words, with respect to response to treatment, chance of recurrence, and overall survival. Indeed, for invasive carcinoma of the breast, grade and stage are currently far more predictive of mortality than (as yet) any molecular markers; such information is more widely used than any measure in diagnostic or clinical contexts (see, e.g., Elston and Ellis,1991; American Joint Committee on Cancer, 2002; Senkus et al., 2015). In breast cancer, the status of axillary lymph nodes (one of the three features identified in the TNM category of stages) is the "most important single prognostic factor for all except a small number of breast carcinomas" (Stankov et al. 2012). The status of axillary lymph node involvement in breast cancer is taken to indicate whether cancer cells have moved outside of the primary site. The more lymph nodes are involved, the lower are the chances of disease-free survival after treatment. To be clear, my claim here is not that "stages" of cancer—such as stage I and stage II— are "fence posts" in nature, in the sense often imagined by philosophers. Rather, the claim is that they track natural facts about the extent of invasion of disease, which are predictive and explanatory of subsequent disease course. Predicting disease course, even in the best circumstances, with the most information, however, is difficult (Leong and Zhuang, 2011). Cancer may move quickly or slowly. Indeed, some cancers fail to progress, and some even regress (though this is controversial; see, e.g., Heim and Köbele 1995).

Stages draw somewhat conventional boundaries in an ongoing natural process; nonetheless, they provide meaningful information about how a cancer is likely to behave. By analogy, we might say that organisms in early stages of speciation can be similarly classified as meeting various conditions intermediate between reproductive isolation and membership in the same reproductive group. Dobzhansky (1937) classified early stages and varieties of modes of speciation (the process by which species diverge and become reproductively isolated groups)[5] in virtue of the acquisition of prezygotic and postzygotic isolating mechanisms.

Subtypes may also be classified in light of distinct gene expression patterns associated with specific mutations. One might think that cancers originating in a specific organ and cell type share the same group of mutations. But this is not the case. While some cancers are genetically "self-similar" to a high degree (e.g., 79% of thyroid cancers have the greatest similarity to other thyroid tumors, according to Heim et al., 2014), some cancers found in the *same anatomic site* are highly *dissimilar* (only 30% of squamous cell lung cancers are genetically akin to other squamous cell lung cancers). Interestingly, as many as 43% of tumors were found to show the greatest genetic resemblance to tumors originating in an anatomic site *different* from their own (Heim et al., 2014, 2363). In other words, about half of the time, cancers originating in different anatomical sites are more similar (genetically) than cancers originating in the same anatomical site, so that a

---

5. To be sure, definitions of "species" are hugely contested (see, e.g., Wilson, 1999; Sterelny and Griffiths, 1999; Ereshefsky, 1992, 1998, 2001).

classification based exclusively on genetics would yield a cross-cutting classification of cancer, at least by the lights of site of origin.

Advances in molecular medicine have suggested to many researchers that "traditional" histopathological classifications based on location, stage, grade, and so on will be replaced by genetic information. However, Kreeger and Lauffenberger argue that

> the relationship between genomic information per se and malignant disease is more elusive than originally hoped . . . even when a specific gene is identified to make a substantial contribution to pathology, that determination does not easily lead to an effective avenue for treatment because of the complex consequences propagated down transcriptional, translational and posttranslational circuits. (2010, 2)

In other words, genetic information (alone) is so far not sufficient to predict the course of disease or to give us all the information necessary for successful treatment. Genetic and molecular data may be used as a *supplement, in order to yield more precise information* about the likely course of disease or the likely response to various targeted treatments. But context matters. That is, *where, when, and in what combination* mutations occur has an impact on how specific mutations affect disease progression and, indeed, how likely it is that various drugs will be effective. The question at issue among researchers today is not whether we ought to replace "traditional" classifications with "modern" molecular classifications, but rather *how best to integrate the variety of information we have, as well as which information is of greater use for one or another particular purpose.*

One proposal for classification that has been linked to prospects for targeted treatments across organ sites is to classify cancers in light of shared aberrations from distinct *pathways*. For instance, though originating in different anatomical sites, some cancers share disruption of families of signaling "networks" associated with specific families of mutations. These different cancers may be similarly vulnerable to targeted therapies. However, attempts to establish that the same drug works in the same way across cancers originating in different tissue types have had mixed success (Turski et al., 2016). This is perhaps not surprising: getting a drug to target in complex organs with a variety of tissues and cell types is different from doing so when the tissue is relatively uniform. Indeed, cancers with aberrations similar to either similar genes or similar networks may behave differently for a variety of reasons. I will mention three.

First, focusing on cell-intrinsic differences, different cancers have distinct combinations of genes mutated: there are usually several pathways affected in cancer. Different cancers with the "same" pathway affected may thus "find" ways to bypass targeted drugs or evolve novel adaptations to targeted therapies, given their distinctive genetic complement (Holohan et al., 2013). Thus, combination therapies—therapies that target more than one pathway—are required. As mentioned earlier, the optimism surrounding classifications based exclusively on the disruption of common signaling networks has been tempered by the lesser effectiveness of drugs across different anatomical sites. This is in part due to the fact that many different pathways are affected in each cancer type and subtype (Turski et al., 2016).

Second, and complicating this further, is the fact that solid tumors may be more or less genetically heterogeneous. That is, a solid tumor may contain several different "subclonal populations," or lineages of cells with distinctive classes of mutations, each of which are themselves evolving and coevolving (Lipinski et al., 2016; Anderson et al., 2011; Gerlinger et al., 2012, 2014). Cancer scientists call this "intratumor subclonal heterogeneity." Indeed, one might cross-classify cancers based on the extent of intratumor genetic heterogeneity or projections of future intratumor genetic heterogeneity or based on various features that predict this outcome (spatial structure, size, mutation rates, and chromosomal instability). This information is important, because the extent of intratumor genetic heterogeneity predicts response to multidrug therapy. In other words, different kinds of genetic information may lead one to classify cancers in different ways in service of different purposes or lead one to cross-classify the same cancers.

Third, each cancer grows in a distinctive tissue microenvironment and has different stromal features, which in turn contribute to the likely response to treatment, risk of invasion, and metastasis. The tumor stroma is the complex of non-cancer cells and non-cellular components that make up a tumor. As much as 80% of a tumor may be composed of stromal components: fibroblasts, immune cells (macrophages, etc.), and the extracellular matrix (ECM), composed of collagens, growth factors, and so on. Cancer-associated fibroblasts (CAFs) can affect many aspects of the tumor, including growth, survival, metastasis, angiogenesis, and immunosurveillance (Liu et al., 2016). Pancreatic ductal adenocarcinoma (PDAC) is an example of a kind of cancer with stromal features that affect prognosis and response to various forms of treatment. PDACs are desmoplastic, which refers to the extent of fibrous or connective tissue occurring around a neoplasm, leading to a dense, "stiff" structure. This feature of PDACs was thought possibly to blunt effective intratumoral drug delivery (Provenzano et al., 2012) and possibly also harbor inflammatory cells with the potential to suppress cancer-directed immune mechanisms. More aggressive PDACs appeared to be more desmoplastic. In theory, therefore, targeting the stroma was thought possibly to enhance drug delivery to the cancer cells. Unfortunately, Provenzano et al. (2012) found that targeting the stroma results in undifferentiated, aggressive pancreatic cancer, uncovering a protective role by stroma in this cancer. However, despite this downside, such cancers appeared also to respond more efficiently to "checkpoint therapies," or therapies that block specific stages of meiosis. In other words, the stroma appeared to be playing a "twofold" role—the stroma might be protective, but removing it might enable more effective directed therapies. In sum, cell-extrinsic factors are relevant to the course of disease and response to treatment (Özdemir et al., 2014).

This does not exhaust the ways in which classifications in light of genetic and molecular features are complicated by tissue microenvironment. Nor does it exhaust the variety of modes of cancer classification. Cancer researchers also classify cancers as either "sporadic" or "familial," where most cancers are the former. The latter "cancer syndromes" are cancers due in part to an inherited predisposition. For instance, Li-Fraumeni syndrome (LFS) is a cancer predisposition associated with germ-line mutations of the *p53* tumor suppressor gene. Patients with this mutation tend to have accelerated age of onset of cancers (cancers arrive at earlier ages), as well as more frequent cancers across a variety of anatomical sites. Similarly, inherited mutations to *BRCA1* and *BRCA2* are associated with higher rates of breast and ovarian cancer. So two instances of the "same" breast cancer, at least as assessed via shared anatomical site, organ, and tissue, as well as cell type and stage of advancement, could be viewed as distinct cancer types, since one is due to an inherited predisposition and another is not. The former are likely to have unique suites of mutations. Indeed, different kinds of germ-line mutations to the "same" genes can yield more or less aggressive forms of cancer (see Malkin, 2017). That is, rather than a single type of mutation, there are several different variants, for instance, of inherited forms of *BRCA1* and *BRCA2*, each with different associated risk of early onset and each at distinct sites (e.g., one is associated primarily with tumors of the breast, the other with both breast and ovarian cancer). So even within a subtype of "familial" forms of breast cancer, one might identify further subtypes, based in part on distinctive deletions, duplications, base pair changes, and so on to families of genes with similar functional roles.

Cross-cutting classifications of cancers can also be generated in light of the particular age or sex of the patient (childhood leukemia, male breast cancer). And cancers may be classified via remote etiology. For instance, smoking-induced cancers are distinctive in their genetic profile and behavior (Alexandrov et al., 2016), as are cancers due to infectious agents (cervical cancers induced by human papillomavirus [HPV], AIDS-associated malignancies). In fact, there are even cancers taken to be strongly associated with pregnancy—some breast cancers seem to be associated with the surge of hormones and changes in breast tissue associated with lactation (Schedin, 2006). Last but not least, cancer systems scientists classify cancers in light of their distinctive topological structure or network features of signaling pathways, and even the geometry of growth planes in solid tumors (see, e.g., Barillot et al., 2013). Such features predict and explain outcomes of interest to both clinicians and basic science researchers, such as the extent and character of profusion of blood supply, which in turn contributes to the "health" or survival and replication of cells in a tumor, and also the chance of successful metastases.

In sum, there are both hierarchical and cross-cutting classifications of cancer types and subtypes, where these classifications arise out of a variety of features of cancers, including but not limited to cell-intrinsic features. There are two general kinds of response one might have to this heterogeneity of cancer classification schemes—the optimistic and the pessimistic. On the one hand, one might think that the science is simply incomplete: we need only do more high-throughput sequencing or, perhaps, further analysis of the data we already have collected, and the answer to how many different kinds of cancer there are will simply emerge. Just as the human genome project has led us to the conclusion that there are twenty thousand human genes, we may simply come to discover that cancer is exactly three hundred different diseases. On the other hand, this same heterogeneity may lead us to the conclusion that there is no correct answer to be found. Perhaps cancer classifications are not genuinely tracking "natural" kinds, but "merely" conventional categories that track our particular interests and purposes.

In my view, both perspectives are misguided. Indeed, the case of the count of "genes" in the human genome serves as a useful object lesson as to why. The claim that there are twenty thousand genes in the human genome depends both on natural facts about the human genome and on a set of decisions made by the community of scientists about what to count (Baetu, 2012). There are different ways of slicing up the human genome and classifying genes, for different purposes. That such decisions about what to label "genes" are conventional[6] does not show that the claim that there are twenty thousand genes is meaningless, however. Once we decide what is being counted (which, however, is no straightforward matter), and why, we may take this number to suggest an interesting set of further scientific questions (why only twenty thousand as opposed to thirty thousand?). Once we have fixed on a particular purpose, we adopt one classification because we have good evidence that this classification is of relevance to our purpose. Such classifications are not arbitrary. Likewise, once a gene concept is specified, it can be an objective matter as to how many kinds of genes there are, and this in turn can become a topic of inquiry.

---

6. Here, by "conventional decisions" I mean decisions where more than one choice is possible, but not all choices may be equally good. Another meaning of "conventional" is that there are different possible choices of convention, each equally good. In my view, once a purpose or goal is fixed, there may well be one or a range of best classification schemes, depending upon how we specify that purpose.

Similarly, we cannot expect a single correct answer to the question of exactly how many cancers there are to simply emerge from more data mining or cancer genomics. The choice to focus on genetic features of cancer is, after all, a choice. Focus on cancer genomics is likely to give us insight into a variety of important aspects of cancer diversity, but not all potential outcomes of interest. In other words, in any case of classification, we face a problem of underdetermination, but this is not to say that once we have made our motivations and sorting principles clear, we cannot count up the kinds in some domain.[7] That our decisions are purpose-relative and empirically underdetermined does not make them arbitrary.

This point is often forgotten; arguably, the confusion goes back to the very author of the concept of a "real Kind."[8] Mill contrasts natural with conventional classifications as follows: "In so far as a natural classification is grounded on real Kinds, its groups are certainly not conventional; it is perfectly true that they do not depend upon an arbitrary choice of the naturalist" (Mill, 1843, 720). Mill appears to be making the claim that a classification is "arbitrary" whenever it is conventional. As Bird glosses Mill, a conventional classification is one not "mandated by the relevant empirical facts about natural similarity and difference" (2015, 3). But empirical facts about natural similarity and difference (whatever this means) do not and *cannot* mandate one classification versus another, at least not in the case of cancer. There are far too many respects in which cancers are similar and in which they are different (cf. Goodman, 1955). The key term in Bird's gloss on Mill is "relevant." Which empirical facts about similarity and difference are the relevant ones? If you think that there is one true classification of "real Kinds" in nature, then you think that what constitutes the relevant similarities is a singular natural fact to be discovered. Anything other than the identification of the similarities and differences of relevance to the one true classification will result in a merely "arbitrary" classification. But this is to make a serious presupposition. Perhaps there are several such relevant similarities and thus classifications; they may lack the features of Mill's "real Kinds," but they may well count as legitimate natural groups.[9]

In order to understand this idea of natural groups, as opposed to merely "arbitrary" classifications, it's helpful to consider the difference between how a child and how a scientist might classify a pile of rocks. The child might classify the rocks on

---

7. It turns out that something very like this view is defended by Ereshefsky and Reydon (2015). I return to this view later.

8. Mill did not use the expression "natural kind." Rather, he spoke of "real Kinds."

9. As Magnus (2015) has argued, Mill's concept of Kinds does not track our current concept. In fact, Mill distinguished natural groups and Kinds. The latter are a subset of the former, where the former, Magnus argues, are categories "that would figure in the science of a disinterested enquirer" (274). Magnus takes Mill to be endorsing "natural groups" as a wider category than kinds, one that would be identified by any ideal, neutral observer, but "natural groups" are not, like "real Kinds," entities with essences and indefinite shared properties. Magnus cites Mill:

the basis of whether the child happened upon the rocks on a Tuesday or any other day. Scientists (in theory) prioritize some similarities over others in a more principled way. The principles are derived (though not deduced) from their background beliefs and scientific theories, as well as their interests in answering specific empirical questions. These interests are (usually) not as idiosyncratic as the child's. Nonetheless, different scientists often have different interests. So we ought not be surprised that they sometimes generate quite different classifications of the same group of entities. For instance, a geologist could well classify a pile of rocks very differently from a chemist. It doesn't follow, however, that their classifications are not in some sense "natural" groups or that they are "merely arbitrary." Geologists and chemists might arrive at overlapping classifications, but they may not. Bursten has demonstrated that "classification in nanoscience differs from classification in chemistry because the latter relies heavily on compositional identity, whereas the former must consider additional properties, namely, size, shape, and surface chemistry" (2016, 1). She uses this case study to argue for a "scale-dependent theory of scientific classification."

In sum, there may well be a number of different yet equally relevant empirical facts about natural similarities and differences between entities, though they may be relevant to different kinds of scientific questions about different (but equally "natural") outcomes. Depending upon which outcome or process we are concerned with, different kinds of similarities, characterized at a variety of grains, will be relevant to our classificatory scheme. Classifications useful for various *predictions* may not be terribly useful for *retrodictions* (historical sciences—cosmology, geology, evolutionary biology—often make claims about the history of the universe or life on our planet). These classifications may not all be "grounded" in the same way. Exactly because scientists' questions are different, we should expect their classificatory schemes—even of the same class of phenomena—to be different. If we are interested in the dynamics of cancer progression, we will likely focus on different features shared among a group of cancers than if we are interested in these same cancers' remote etiology. Of course, sometimes one classificatory scheme will be useful or informative in service of addressing questions it was not intended to answer, but not always. Philosophers have (mistakenly, in my view) tended to assume that unless a scheme is univocal—or applies across a variety of different contexts, the scheme does not reflect "natural" classifications; but it is unclear why this must be the case.

For example, consider the scientific question of evolvability: What enables a population of organisms or a lineage to evolve? There are several ways in which this

---

[W]hen we are studying objects *not for any special practical end*, but for the sake of extending our knowledge of the whole of their properties and relations, we must consider as the most important attributes those which . . . would most impress the attention of a spectator who knew all their properties by was not especially interested in any. Classes formed on this principle may be called, in a more emphatic manner than any others, natural groups. (Mill, 1874, 500–501)

property is realized; or, if you like, there are several senses in which populations are more or less "evolvable." The heterogeneity of a population is a "natural" fact of relevance to a "natural process," namely evolution by natural selection. Broadly speaking, heritable variation is a necessary condition of natural selection, and the more such variation is available for selection to act upon (all else being equal), the more quickly a population will respond to selection. Populations with more rather than less heritable variation are thus in a sense highly "evolvable." This definition of evolvability is useful for the purposes of questions about evolvability, or adaptive change, within populations. But suppose the outcome of interest to a group of evolutionary biologists is not within population evolvability, but relative evolvability of lineages. Some lineages are relatively diverse—a nice example here is beetles, of which there are 350,000 species described (with likely more to be discovered). Lineages (as opposed to populations) might be more or less "evolvable" in virtue of their possession of not only heritable variation at the population level, but also lineage-level shared features like modularity of organization (see, e.g., Pigliucci, 2008). Scientists (arguably) meaningfully characterize both populations and lineages as "evolvable" in virtue of their possession of similar properties (multiply realized) that predict the disposition of such populations to evolve (Brown, 2013). Indeed, some cancers are described as highly "evolvable" in light of similar features. Let us turn to unpacking the notion of natural kind a bit further, to see how and why it has shaped the debate about natural classification, and classifications of disease kinds in particular.

1.3. Natural Kinds and Disease Kinds

Natural kinds are traditionally[10] understood to be discrete types[11] found in the natural world, discovered rather than invented. For instance, sodium

10. As we've already seen, whose tradition is, to some extent, an open question. Whewell's "natural classes" rather than Mill's "real Kinds" may well have been closer to the sense of "natural kind" that many philosophers seem to endorse. Whewell seems to think that kinds have determinate properties, which belong all and only to each kind, and that these have to do with kinds constitution or microstructure: "since the truths we are to attend to are scientific truths, governed by *precise and homogeneous relations*, we must not found our scientific Classification on casual, indefinite, and unconnected considerations" (Whewell, 1858, 115). And "the science which we require is a complete and consistent classified system of all inorganic bodies. For chemistry proceeds upon the principle that the constitution of a body invariably determines its properties; and consequently, its kind" (Whewell, 1837, 189). See, e.g., Khalidi (2013), Magnus (2014), Snyder (2006), and Hacking (1991, 2007) for an enlightening discussion of the history of the idea of a "natural kind."

11. Metaphysicians debate whether kinds are types of entities or can be reduced to something else—universals, clusters of properties, etc. For discussions of such topics, I recommend Ellis (2001), Lowe (1998), Armstrong (1978, 1997), Hawley and Bird (2011), Boyd (1991), and Millikan (1999).

(Na), radon (Rn), and helium (He) are said to be natural kinds. The identification and classification of natural kinds is often taken to be a central aim of science. Theoretical terms—terms that figure centrally in scientific laws and theories—are said to pick out natural properties, and also natural kinds. Kind predicates ("gold," "mercury") are said to be "projectable" in light of the shared properties of entities that fall under these kind categories. For instance, what makes gold a kind, on this view, is not its observable features (yellow color, malleability, ability to conduct electricity) but the physical structure belonging only to gold—that is, its distinctive microstructure. These shared intrinsic features explain and justify our expectations that kind tokens behave similarly in similar instances. For instance, electrons are all negatively charged subatomic particles approximately 1/1836th the mass of a proton; and all samples of potassium are metal, with nineteen protons in their atomic nucleus, oxidize in air, and are reactive with water.

Disease classifications (nosologies) may at first appear to play just such a role in medicine: they help organize common disease types, unify suites of symptoms, and aid in diagnosis and prognosis. Expecting that, for instance, several individuals with common symptoms will have a common response to a course of treatment, or a common course of illness in the absence of treatment, appears to be the foundation of medicine as an inductive practice. The presence of some underlying intrinsic state or process, one might think, should explain shared signs and symptoms. It would seem that a *realist* about medicine *as a science* must be committed to the existence of objective disease kinds and objective bases of kind classifications; otherwise, the success of medical practice might appear to be miraculous (cf. Williams, 2011).

Indeed, many historians trace the rise of medicine as a science to the adoption of this model of disease as distinct biomedical kinds. In the late nineteenth century, some of the first advocates of the germ theory of disease, such as Semmelweiss and Pasteur and later Koch (who developed "Koch's postulates"),[12] argued that specific pathogens or infectious agents could be isolated to determine exactly how they led to common symptoms and diseases in different patients. Greene summarizes a modern version of this biomedical model of disease categories as follows:

> Part of the power of biomedical categories, both diagnostic and therapeutic, is their apparent universality; diseases and cures, abstracted from local context and understood on a microscopic level of molecular mechanism, should in theory work everywhere, should in theory be the same

---

12. For a philosophically sophisticated discussion of Koch's postulates, see Ross and Woodward (2016).

everywhere. This kind of biomedical knowledge claims to be universal. Biomedical objects can be found everywhere and are everywhere expected to perform the same way. (2014, 12–13)

This model may seem appropriate for more clear-cut cases of disease "kinds," where there is one single, well-defined causal agent. For instance, infectious diseases share a causal agent, such as a particular species of virus or bacteria with an intrinsic set of features unique to that species (setting aside for the moment that all viruses and bacteria themselves evolve, and coevolve with hosts). However, many diseases lack a single common cause. "Multifactorial" diseases like cancer, heart disease, and arthritis lack a common cause and so would seem to fail to count as natural kinds on this view of disease kinds (cf. Williams, 2011).

There are several kinds of response one might consider in the face of this complication. First, some have argued that for complex diseases like arthritis and cancer, we ought to adopt a *weaker* standard for kind status. For instance, some contend that such diseases are instances of what Boyd (1999) has called homeostatic property cluster (HPC) kinds (Williams, 2011; Khalidi, 2013). On this view, kinds need only have shared properties that stably cluster in virtue of a shared underlying mechanism or mechanisms. Alternatively, one could argue for a different account of kinds, for example "functional" or perhaps "dysfunctional" kinds (see, e.g., Lange, 2007; Sorensen, 2011).[13] On this latter view, what it is that kinds of disease have in common is similar disruption of specific functions. I will discuss these views in turn. Let us consider the homeostatic property cluster view first.

---

13. To be precise, Sorenson says that we ought to consider the possibility of diseases counting as para-natural kinds, where para-natural kinds "inherit the lawfulness and projectability of the natural kinds that shape them" (2011, 117). In contrast, natural kinds "owe their features to their *internal nature*. They have characteristic origins, and characteristic patterns of change. This regularity is underwritten by a rich network of causal interactions that are kept stable" (117). So on Sorenson's view, para-natural kinds are reflections or patterns in phenomena that in some way depend for their regularity upon the behavior of the more fundamental natural kinds. Shadows serve as an example. Sorensen's reasoning for taking diseases to be para-natural kinds is as follows: "[D]iseases appear to be deficiencies, departures from the norm that compromise survival and reproduction. These privations lack integrity. Pathogens can exist separately from the organisms they infect (and are themselves subject to diseases. But disease itself depends on its sufferer . . . One way to reconcile this dependency with the resemblance to a natural kind is to characterize diseases as para-natural kinds" (122). Sorensen grants that if we accept natural kinds, we have to take both "first-order" and even second- and third-order para-natural kinds on board. By "parity," he means that "what goes for natural kinds goes for para-natural kinds."

On the HPC view, the shared presence of one or more "homeostatic" mechanisms favors the presence of others, and thus ensures the persistence of the cluster, much as metabolic processes (among others) ensure "homeostasis" in the human body. Homeostasis need not be perfect. Properties possessed by some members of an HPC kind need not be possessed by all, and clusters of properties may shift over time. Nonetheless, the existence of shared mechanisms explains the success of our inductive practices. This view was initially proposed to solve the problem of characterizing species as natural kinds, for species seem to pick out genuine kinds in biology but lack "essences." They are heterogeneous, and what's worse, they evolve. Boyd proposed that we ought to think of species as cluster kinds, which are stably maintained in virtue of homeostatic mechanisms—in this case, interbreeding. Several philosophers have echoed the view that the HPC approach is especially suited for biological kinds, such as species, where there are no shared essences but common properties that tend to cluster (e.g., Boyd, 1999) (though several philosophers have taken issue with this account of kinds, most recently Ereshefsky and Reydon, 2015).

Since diseases are likewise demarcated via the identification of sets of symptoms, properties, or dispositions that stably cluster, or occur together, diseases might count as HPC kinds. What causes such properties to cluster may not be a single mechanism shared by all instances of such kinds, but instead a variety of mechanisms, none of which are uniquely necessary or sufficient for disease membership (Williams, 2011). The advantage of this view is that it accommodates both the heterogeneity of causal bases of diseases like arthritis and cancer and the fact that such kinds are fuzzy and relatively "polythetic": that is, not all properties are shared, or shared to the same extent. It seems that cancer would count as a vivid instance of just this kind of kind.

1.4. Is Cancer a Homeostatic Property Cluster Kind? A Potted History and Discussion

A bit of historical perspective may be of value in addressing this question. "Cancer" has not always been viewed as a single disease. However, several "general" theories of cancer have been proposed. One theory identifies cancer as essentially a "genetic," another as a "viral disease," another as an "environmental" disease. Each has been false; or perhaps it is better to say that each has been only partially right. For instance, Virchow (1870) held that cancer was a product of chronic irritation or inflammation. He was correct (in part); *some* cancers are indeed strongly associated with long-term inflammation, and an inflammatory environment can promote cancer. In the 1960s, work on the Rous sarcoma virus, as well as the discovery of the role of HPV in cervical cancer, lent hope to the idea that all cancers are caused by viral infections (Clarke, 2011). Again, this theory is also false; only about one-fifth of cancers are induced by viral infections. What of the theory that cancer is an "environmental" disease? Again, this is only true (in part): decades of careful epidemiological work have demonstrated that environmental exposure to various carcinogens does cause cancer. Pott discovered that rates of scrotal cancer among chimney sweeps were high, Rehn found a high incidence of bladder cancer among workers in the German synthetic aromatic amine dye industry, and Wynder and Graham, as well as Doll and Hill, linked lung cancer with smoking. However, environmental causes can account for only about one-third of cancer mortality, though they likely contribute to a significant portion of cancer incidence (see Chapter 4).

One might say that all these theories were in error because they focused on remote rather than proximate causes of cancer; that is, they failed to identify the "mechanism" by which inflammation, infection, or environmental exposure caused cancer. Such an argument is what has led us down the garden path to the view that cancer is a "genetic" disease, or a disease due to mutations of cells, mutations that play key roles in the regulation of cell birth and death. This view has, of course, massive support and a great deal of rhetorical power. The narrative history of this view often starts in 1971, though it is much older. In 1971, Alfred Knudson, a pediatric oncologist, noticed a pattern. He was studying retinoblastoma, a rare form of childhood cancer where tumors appear in the retina in one or both eyes of patients (called "unilateral" and "bilateral" cases, respectively). Knudson noticed that children with tumors in both eyes seemed to present in the clinic much earlier than children with only one eye affected. What could explain this difference? Knudson was familiar with family histories of retinoblastoma, which seemed to show that the gene was inherited in a dominant fashion; that is, 50% of children of an affected parent were likely to be affected. He was also familiar with epidemiological evidence suggesting that cancer incidence by and large increased as a function of age.

In the 1950s, Nordling (1953) proposed that the age dependence of cancer incidence might be explained by the acquisition of a series of mutations over the course of a lifetime. Two epidemiologists, Armitage and Doll (1954), likewise offered a "multistage" theory. They noticed an age-dependent incidence of lung cancer; and it appeared that smoking might "shift" the curve for the age of incidence, so that smokers presented with cancer at earlier ages than other cancer patients and (of course) far more frequently. Knudson drew upon all this evidence and generated a very precise, testable hypothesis: "[R]etinoblastoma is a cancer caused by two mutational events. In the dominantly inherited form, one mutation is inherited via the germinal cells and the second occurs in the somatic cells. In the nonhereditary form, both mutations occur in somatic cells" (1971, 820). This hypothesis (sometimes called the "two-hit" theory) predicted the patterns of incidence in bilateral versus unilateral cancers that Knudson saw in the clinic. Knudson's prediction was, in a sense, borne out.[14] *RB* (the retinoblastoma gene) was later found to be inherited in families with a high incidence of bilateral tumors; *RB* was the first identified tumor "suppressor" gene, a gene ordinarily associated with halting the advance of the cell cycle. Knudson, however, knew nothing of the mechanistic bases of this inherited mutation; he developed this hypothesis exclusively on the basis of observed patterns of incidence. Earlier onset and more devastating cases were, he suspected, due to inherited mutations; these children needed to pass through fewer steps to get cancer than sporadic cases.

Successes like Knudson's have had a profound effect on our current understanding of cancer. The story of the two-hit theory, its precise and successful predictions, and the subsequent discovery of the *RB* gene makes for exactly the kind of satisfying resolution to a scientific puzzle that can convince researchers and their backers to hunt down yet more and similar causes. In the 1980s, research into the genes "for" cancer, or "proto-oncogenes," became a massive political, financial, and social movement. The successes of the "oncogene paradigm" and the associated technological advances created a "bandwagon" effect on research, which has, over the past twenty-five years, focused research dollars on the cell and molecular biology of cancer or largely cell-intrinsic mechanisms associated with cancer (Fujimura, 1997). Scientists involved in the oncogene bandwagon lobbied Congress and acquired resources from the Food and Drug Administration, the National Cancer Institute, and private industry. That is, cancer genetics has become big business and currently receives a significant portion of federal funding (relative to research, genetic or otherwise, on other diseases). The Cancer Genome Atlas project, launched in 2005, has sequenced over 11,000 of tumor samples and generated the "cancer genome" of about thirty different cancer types. Hundreds of genes have been identified that play an important causal role in cancer. These are sometimes called "oncogenes" and "tumor suppressor" genes—mutations to which advance or promote "hallmark" behaviors of cancer cells (Hanahan and Weinberg, 2000). *SRC, WNT, TP53, APC, RB, BRC-ABL, BRCA1,* and *BRCA2* are a few examples of genes that play a key role in cancer initiation and progression.

This picture of cancer as a "genetic disease" is taught to medical students and also informs our popular understanding of the disease. Current textbooks

---

14. Of course, not all cancers are caused by "two hits," nor is retinoblastoma, strictly speaking. More on this point later.

in the molecular biology of cancer come with poster-sized diagrams of genetic networks, looking rather like diagrams representing electronic circuits. These elaborate connections between genes and their products play a role in "halting" or "promoting" tumor growth. In addition to retinoblastoma, a variety of other "familial cancer syndromes" have been identified. These are cancers that arise in families, and they have been linked to inherited mutations to "tumor suppressor" and "proto-onco-" or "oncogenes"—the very same mutations that also appear in "sporadic" cancers. That is, the same mutations to the same genes occur in both inherited and acquired disease, just as in the case of unilateral and bilateral retinoblastoma.

The success of this view of cancer has to do at least in part with the fact that the multistage theory predicts what we observe: cancer incidence increases (by and large) as we age. If mutation is a rate-limited process, a series of mutations to cells acquired over the course of a lifetime will yield these patterns in incidence, and this is (roughly) what we see. However, mutations to cancer cells alone are by no means sufficient for invasion and metastasis, as we shall see. Nonetheless, this theory has been so successful in part because it provides a unified *and* mechanistic theory of cancer causation, linking molecular genetics, toxicological evidence, and epidemiological evidence in a common theoretical framework (Plutynski, 2013). This has made the theory enormously persuasive, reinforcing the idea that we ought to think of cancer (in general) as a "natural" kind, and has provided cancer scientists with a research program, one linked tightly to a hope for targeted intervention.

Two researchers who have played very prominent roles in this research—Hanahan and Weinberg—offered a general characterization of the "hallmarks" or capacities (perhaps better: "incapacities") of cancer (Hanahan and Weinberg, 2000), which they updated in 2011. Consider the schematic description of six capacities paradigmatically associated with cancer shown in Figure 1.1. According to Hanahan and Weinberg, cancer is caused by failed regulation of cell division, failure of apoptosis, as well as overactivation of pathways ordinarily suppressed: sustained growth signaling and activation of signals that promote angiogenesis, invasion, and ultimately metastasis. All these failures are, they argue, linked primarily to particular mutations in cancer cells. To be clear, the schematic in Figure 1.1 does *not* describe a mechanism in the sense of an organized sequence of events. Rather, the mechanisms that underpin each of these capacities are massively heterogeneous—there is no single mechanism, for instance, for "limitless replicative potential." Moreover, the process is not organized, in the sense that some such capacities are acquired (generally speaking) earlier and some later, though what it means to "acquire such a capacity" is somewhat vague, just as *what it is that bears the capacity* is similarly vague. That is, these are described as "capacities of cancer," but sometimes it seems we are to understand them as

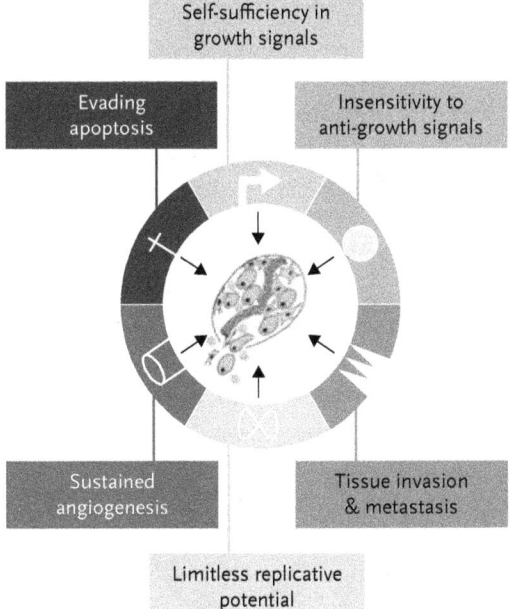

FIGURE 1.1 The hallmarks of cancer—Hanahan and Weinberg's iconic image, representing six "functional capabilities" acquired by "most if not all cancers . . . during their development, albeit through various mechanistic strategies": sustaining proliferative signaling, evading growth suppressors, resisting cell death, enabling replicative immortality, inducing angiogenesis, and activating invasion and metastasis.

From Douglas Hanahan and Harry Weinberg, "Hallmarks of Cancer: The Next Generation," *Cell*, 100, no. 1 (2011), 57–70.

capacities of cancer cells, whereas other processes (e.g., metastasis and invasion) are to be understood as capacities of cell lineages or perhaps whole tumors.

Many such capacities of cancer cells are caused not (or not only) by cell-intrinsic changes (e.g., mutations, epigenetic changes) but also by cell-extrinsic factors. To be sure, some capacities are more "autonomous" than others; for instance, autocrine signaling is the capacity to sustain proliferative signaling by producing growth factor ligands and to respond via the expression of cognate receptors. The acquisition of this capacity can occur simply due to cell-intrinsic changes. But other mechanisms for the same capacity require signals from the surrounding environment. Indeed, many of cancer's capacities require for their activation molecules and other elements of the tissue microenvironment and, indeed, involve the co-option of normally adaptive responses of surrounding tissue. In a 2011 paper, "The Hallmarks of Cancer: The Next Generation," Hanahan and Weinberg are quite explicit on this point: "[T]hese stromal cells contribute to the development and expression of certain hallmark capabilities. During the ensuing decade this notion has been solidified and extended, revealing that the biology of tumors can no longer be understood simply by enumerating the traits of the cancer cells but instead must encompass the contributions of the 'tumor microenvironment' to tumorigenesis" (144). They add several "emerging hallmarks" and "enabling characteristics" to the original six capacities to emphasize this fact, as shown in Figure 1.2.

Note that these properties are not simply properties of cancer cells, but properties that require dynamic interactions between tumor cells and their environment. Nonetheless, the picture that most take away from the hallmark view is that cancer is a "kind," whose properties are driven primarily by changes to cancer cells, specifically mutations. Many cell and molecular biologists still often write that cell-intrinsic mechanisms (alone) drive changes to the tumor microenvironment. Hanahan and Weinberg describe the characteristic genomic instability of cancer cells as "enabling" the acquisition of further mutations, many of which they consider essential to cancer's hallmarks. They describe mutations to "caretaker" genes, which ordinarily play an important role in detecting DNA damage, activating repair machinery, repairing damaged DNA, and activating or intercepting mutagenic molecules, as key enabling causes of cancer, akin to a doorway that, once broken down, permits the acquisition of further mutations.

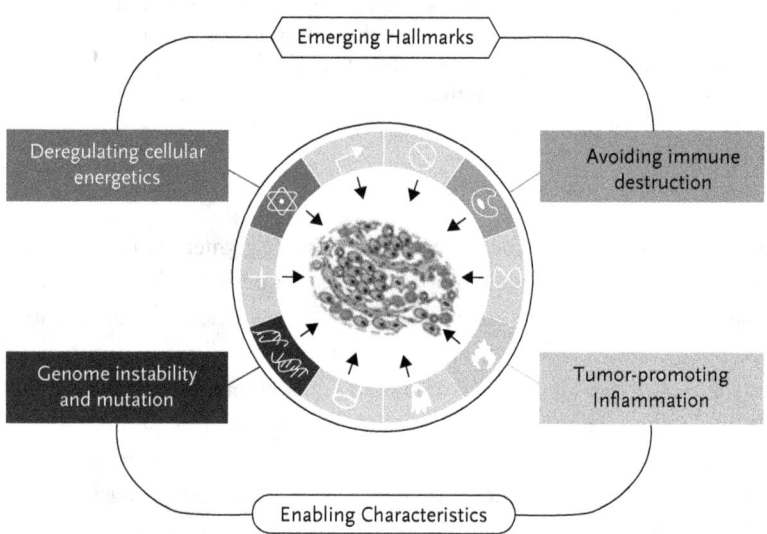

FIGURE 1.2 The emerging hallmarks and enabling characteristics of cancer—Hanahan and Weinberg's expanded list of capabilities "involved in the pathogenesis of some and perhaps all cancers": "the capability to modify, or reprogram, cellular metabolism" and "evade immunological destruction." "Genomic instability" and "inflammation" also enable tumor progression.

From Douglas Hahahan and Harry Weinberg, "Hallmarks of Cancer: The Next Generation," *Cell*, 144, no. 5 (2011), 646–674. © 2011 Elsevier Inc. Published by Elsevier Inc.

Yet almost two decades ago, it was experimentally demonstrated that cancer may be initiated by wounding, infection, or persistent inflammation (Martins-Green et al., 1990, Martins-Green and Hanafusa, 1997; Olumi et al., 1999; Sternlicht et al., 1999; Coussens and Werb, 2002; Maffini et al., 2004). That is, it appears that in at least *some* cases, mutations are not initiating conditions, but may in some cases *follow upon* disruption of normal tissue organization, as critics of the mutation theory attest (see, e.g., Sonnenschein and Soto, 1998).

So is cancer an HPC kind? Khalidi (2013) makes the strongest possible case for the view that cancer (in general) is an HPC kind. According to Khalidi, of all the natural kinds he considers in his book (including, but not limited to, chemical kinds, kinds of mental illness, viruses), cancer "is perhaps the most in line with Boyd's homeostatic property account" (2013, 183). Unlike other biological kinds, he explains, "for which there may be no single well-defined mechanism" (77), Khalidi suggests that "caretaker" mutations qualify as the homeostatic mechanisms for cancer, because they "are the causal mechanism that gives rise to the cluster of other properties associated with cancer cells" (183). These mutations are characteristically found in cancer cells; insofar as they are necessary for cancer,[15] they are the mechanism that defines cancer (in general) as a kind. Since, Khalidi argues, "the kind of entity that is most closely implicated with the process of cancer is the cancer cell . . . once the cancer cell is understood as a natural kind of entity, then cancer can be seen as a natural kind of process" (181). Khalidi's argument can be reconstructed as follows:

1. Homeostatic property cluster kinds are kinds for which there is a "homeostatic mechanism" or mechanisms.
2. *Cancer cells are an HPC kind*; they have shared causal properties, and these properties are all traceable to mutations to caretaker genes.

---

15. Thank you to a kind reviewer for the following information. Whether mutations are "necessary" for cancer is somewhat contested. For instance, a paper by Mack et al. (2014) appears to show that a cancer can in theory develop without a mutation to "cancer cells." In particular, a mutation (in *Ptpn11*) in the bone marrow cells in mice results in the emergence of abnormal hematopoiesis mimicking a particular kind of myeloproliferative neoplasm called juvenile myelomonocytic leukemia (JMML; characterized by an excess of monocytes). That is, there is an absence of mutation in the leukemic cells themselves; only the cells in the tumor microenvironment are mutated (*Ptpn11* mutation). There is some debate about whether this is really a cancer, which is due in part to the model and in part to the intrinsic limit of hematological malignancies. (There is no metastasis for those kinds of cancer, so whether reproducing the excess of monocytes can count as a case of JMML or CMML [the adult version of JMML] is a difficult question. See Dong et al., 2016). Another notable exception, found not in leukemia but in ependymoma, is a tumor of the central nervous system that can be benign or malignant. Mack et al. (2014) sequenced by whole exome or whole genome a number of these tumors and found very few mutations. In one patient, they found no mutations at all. However, one should be very cautious in interpreting the data.

3. Such mutations function as "homeostatic mechanisms," in that they induce further mutations, which in turn yield the persistent traits typical of cancer cells, or the *hallmarks of cancer*—such as replicative immortality, resisting cell death, and evading growth suppressors.
4. If cancer cells are an HPC kind, *then cancer is a natural kind*.
5. Thus, cancer is a natural kind.

I have three major concerns with this line of argument.[16] First, it is unclear whether cancer cells are an HPC kind. Second, it is not clear whether caretaker mutations per se are best described as "mechanisms for" cancer. Third, even if we were to establish that cancer cells are kinds, it would not necessarily follow that cancer is a natural kind. Whether cancer *cells* are kinds and whether *cancer* is a kind are different questions; answering the first question does not (necessarily) help us answer the second.

Let us turn to the first objection. Are cancer cells a "natural kind"? What makes something a "cancer cell"? Perhaps cancer cells are any cells with features that enable them to contribute to disorderly growth or invasive behavior. This definition is too permissive, however; it would require "cancer cell" to refer to almost any cell in the body involved in some way or another in assisting the growth of a tumor; but cells in the heart and lungs assist in providing a blood supply to a tumor. Indeed, there are even specialized "cancer-associated" cells in the stroma, or supportive tissue surrounding a tumor, that contribute to disorderly growth and invasive behavior. These include cancer-associated fibroblasts (CAFs) and myofibroblasts, which help degrade the extracellular matrix and enable invasion; endothelial precursor cells (EPCs) from the blood marrow that differentiate into blood vessels and platelets, and release platelet-derived growth factor (PDGF), which increases the permeability of blood vessels and attracts fibroblasts, monocytes, and neutrophils, the latter in turn activating the growth of cells and assisting in angiogenesis; and matrix metalloproteinases (MMPs), which play a role in remodeling the extracellular matrix and release factors that stimulate and permit the epithelial–mesenchymal transition (EMT), imparting motility and invasiveness to metastatic cells. Indeed, as many as "90% of the cells in a tumor mass" in breast, colon, and stomach cancer may be made up of a complex mix of these "non-neoplastic" cells (Weinberg, 2007, 529), which play a variety of cancer-promoting roles. If any cell that contributes to or promotes the

---

16. To be clear, however, my aim here is not to criticize Khalidi (with whose views on most matters regarding kinds I am very sympathetic) so much as to use his discussion as a way to explore the scope and limits of the HPC view of kinds. That is, my aim here is to uncover the intrinsic challenges in applying the HPC approach.

growth of a cancer is a cancer cell, then we ought to conclude that all such cells are "cancer cells."

Second, it is worth noting that cell biologists deploy different criteria for cancer cells in vivo (in the body) than those considered appropriate for identifying cancer cells in vitro (in culture). This is because the properties of such cells are different, and this difference is in part a product of design. Cancer cells in the body lack many of the features considered typical of cancer cells in culture, or "cancer cell lines." Cancer cell lines are "immortal" (i.e., they can be passaged through culture indefinitely), they grow without anchorage dependence (they do not depend upon an extracellular matrix to grow and divide), and they should produce tumors when injected into experimental animals. But most cancer cells in vivo do not meet these conditions, or won't without a good deal of manipulation. It took decades of careful work to grow a cancer cell line outside the body (Skloot, 2010). Most cells taken from tumor biopsies lack the special features typical of cancer cell lines, such as HELA cells. Primary cultures of cancer cells (drawn from a tumor) will grow in culture, but they rarely grow indefinitely (i.e., they are not "immortal").

One might respond that cancer cell lines such as HELA are artifacts, or at best hybrid kinds—entities that may well have been manipulated to serve various experimental purposes.[17] Since our question concerns whether cancer cells are natural kinds, it is not clear whether the properties of these artifacts are relevant to our argument. However, most experimental work on cancer is premised on the idea that the properties and behaviors of cancer cell lines are representative of the natural kind, "cancer cell." Why? Presumably, it is those properties *shared* by cancer cells in vivo and cancer cell lines that are "essential" to something being a cancer cell. What are those features?

One central feature is that both kinds of cells typically have increased expression of "oncogene proteins." For instance, mitogens (growth factors) are typically produced in excess by cancer cells, because the pathways that control mitogen production or uptake are disrupted. Is the expression of mitogens sufficient to count cancer cells as distinct kinds? The answer is no; for, mitogens are, as a matter of fact, expressed by a variety of cells over the course of development. Many cells (in the colon, skin, esophagus, etc.) are constantly dividing and shedding, and this process requires mitogens. In other words, it is possible for "non-cancer" cells to express "oncogene proteins." The difference is a matter of *degree*, not of kind. It seems that what is distinct about cancer cells is not the expression of mitogens per se, but the failure to regulate their expression in an appropriate way.

Nor is it clear that the possession of many "cancer" mutations—or mutations underlying cancer's hallmark capacities—is sufficient for a cell to be characterized

---

17. For an intriguing discussion of cells as technologies, see, e.g., Landecker (2007).

as a cancer cell. It turns out that many cells found on the surface of the skin or the lining of the gut possess these mutations (and thus, at least potentially, capacities). Martincorena and colleagues (2015) found a large number of "driver" mutations in cells taken from normal sun-exposed adult skin (the eyelids of patients who had lifts to their eyes). These were mutations associated with hallmarks of cancer: "there were more NOTCH1 mutations in just 5 cm² of aged, sun-exposed skin analyzed here than have been identified in more than 5000 cancers . . . we found clones carrying two to three driver mutations that had not acquired malignant potential, raising the question of what combinations of events are sufficient for transformation" (Martincorena et al., 2015, 885). That is, *one-fifth of healthy skin cells* contain as many mutations as is typical of cells found in a tumor.

It seems that whether we call cells with mutations to "cancer genes" "cancer cells" depends on whether we happen to find them in a cancer. Or to be a bit less cynical, it seems that mutations, which are components of the "mechanisms" that cause cancer cells to behave as they do, are not unique to or sufficient for cancer. Nor are such mutations sufficient for "homeostatic" maintenance of cancer phenotype. Indeed, cells taken from a tumor have been shown to "revert" and become part of a healthy tissue in a developing embryo, given the appropriate signals from surrounding tissue (cf. Bissell and Hines, 2011). For most of the lifetime of the organism, the mechanisms that prevent cancer are, by and large, "more homeostatic" than the mechanisms that (eventually, as we age) lead to cancer. Otherwise, we would get a lot more cancer than we do (Bissell and Hines, 2011).

This may seem surprising at first pass, but not once we consider our evolutionary history and how it has shaped the mechanisms that by and large prevent cancer. Consider the tissue architecture of skin and the basic biology of cell renewal. Over the course of the lifetime of the average individual, epithelial tissue (such as the lining of the skin, esophagus, and gut) is constantly being sloughed off. Epithelial tissue in the colon, for instance, has a hierarchical structure, where cells migrate to the surface and die. Somatic stem cells reside at the base of "crypts," whereas differentiated cells migrate to the surface and are shed. The cycle time is five to seven days for a given cell; thus, the entire epithelium of the gut is turned over roughly every seven days. Many thousands of cell divisions are ongoing throughout the body, and randomly distributed somatic mutations occur and accumulate in such cells on the surface of the skin. Most cells that acquire these mutations do *not* progress to invasive cancer; there are, in fact, many "homeostatic mechanisms" that act to *prevent* cancer. As Martincorena et al. explain, "For cancers to occur with the frequency they do in the general populations, there must be a vast *underlying reservoir of competing clones part of the way to malignant transformation*" (2015, 885, emphasis added). So the characteristics of cancer cells are (a) not unique to cancer cells and (b) only more or less homeostatic. This should not surprise us; the suite of mechanisms that (for the most part) prevent the growth and development of invasive disease have had millions of years to evolve this far more "homeostatic" capacity.

Perhaps we ought to narrow down the definition of cancer cells to all and only cells that possess both mutations to cancer genes, produce oncoproteins, and, in fact, participate in or contribute to invasive disease, i.e., cancer. However, this definition faces a yet more serious difficulty. For recall that according to the argument above, the aim was to characterize cancer as an HPC kind by first establishing that cancer cells are an HPC kind. That is, Khalidi's argument relies on the assumption that we should be able to arrive at an account of cancer *cells* as HPC kinds—one that does not depend on a prior account of *cancer* as a kind. If we were to rely upon "invasiveness" in defining "cancer cells," then the account would be circular. Moreover, as we've seen, when and how something functions "as" a cancer cell often depends upon the context in which it finds itself; indeed, this may well be true for most cell types.[18] Tissue-specific traits are often lost in culture, because the regulation of function in vivo depends on complex recip-rocal interactions between cells and their microenvironment. It is these complex processes of reciprocal interaction that define "cell types"—indeed, as discussed earlier, a breast cell or prostate cell can be transferred into an embryo and develop into normal tissue. It is the dynamic reciprocity of this process that determines cell function and differentiation—differentiation is not "fixed" forever (see, e.g., Bissell et al., 1982; Bissell and Aggeler, 1987; Nelson and Bissell, 2005, 2006; Xu et al., 2009). One cancer researcher, when asked what, apart from binding with various antigens, typifies healthy versus cancer cells, told me that it is their dis-tinctive *interactions* with neighboring cells. Healthy cells "dance" together like expert salsa dancers, he explained, whereas cancer cells behave like dancers at a Grateful Dead concert—they "go their own way" (Muhoro, 2017).

Turning to the second objection, are caretaker mutations properly viewed as "mechanisms for" cancer? In order to address this question, we must first consider what counts as a mechanism. There are a variety of competing views on "mechanism" (see, e.g., Machamer et al., 2000; Glennan, 2002 Bechtel and Abrahamson, 2005; Craver, 2007), but they share a set of features (to which

---

18. Bissell (1981) makes this point rather forcefully. She argues that attempts to identify de-fining features of "normal" cells, let alone "cancer cells," are confounded by the fact that cells behave differently in different environments. Since their behavior varies so significantly in culture, cell biologists searching for the "essential" features of either "differentiated" or "de-differentiated" cells of various types run into a suite of difficulties. In her words:

If there is one generalization that can be made from all the tissue and cell culture studies with regard to the differentiated state, it is this: since most, if not all, functions are changed in culture, quantitatively or qualitatively, there is little or no "constitutive" regulation in higher organisms; i.e., the differentiated state of normal cells is unstable and the environment regulates gene expression. (1981, 27).

we will return, in greater detail, in Chapter 3): they are characterized as either "structures" or "systems" composed of parts, entities, and processes, and organized so as to produce a given behavior, perform some specific function, or yield a specific output, given specific inputs or initial conditions. According to one widely cited view, a mechanism is a set of "entities and activities organized in such a way that they are productive of regular changes from start or set-up to finish conditions" (Machamer et al., 2000, 3).

But even this relatively precise definition leaves the characterization of "mechanisms for cancer" relatively open-ended. What counts as a start-up condition or finish condition? In cancer, this is far from obvious, to say the least. What we choose to call a mechanism may depend upon how permissive we choose to be; there are more or less "paradigmatic" mechanisms and mechanistic explanations. As Woodward (2013) argues, the paradigmatic "mechanism" has properties or features typical of a machine, such as a Rube Goldberg apparatus. It has discrete, organized, modular parts, each of which stably reproduce the same outcome, given the appropriate, fine-tuned input and details of spatial and temporal organization. While mutations may be one of several *events* properly characterized as a *part* of one of many mechanisms for cancer, they are not themselves mechanisms. Mutations are neither organized nor decomposable into parts and activities; nor do they perform a (discrete) function, except insofar as they are enabled to do so via the activity of other entities in their intra- and extracellular context. The pathways associated with "cancer" genes are often coregulated with other pathways. For instance, the same pathway may halt the initiation of apoptosis (cell death), initiate the growth (or breakdown) of an extracellular matrix (a structure that enables tissues to remain relatively stable), enable cell motility (which enables invasion), or attract fibroblasts (other structural features of some tissues, some of which may be co-opted in cancer development). Many of the same "hub" genes are associated with several of these pathways. That is, they are pleiotropic in their effects, and many of these pathways—when disrupted—can overactivate processes associated with cancer. Such overlapping and interacting causal processes raises the following puzzle: how many mechanisms are there in this set of pathways?

It turns out that drawing the boundaries around a mechanism "for" cancer is a pragmatic process, involving choices about what we are interested in explaining (cf. Craver, 2009). Almost any causal explanation that involves some decomposition of parts and processes, regular start- and setup conditions, and termination conditions could count as a mechanistic explanation. So "mechanisms" for cancer could (in principle) include everything from cell-intrinsic mechanisms to environmental conditions that regularly induce cancer, provided we decompose such processes into entities and activities with some sort of organization and "productive continuity." Whether we wish to say that there are two or two thousand mechanisms for cancer, let alone whether these are homeostatic enough to establish a 'stable' kind, may well depend upon how we carve up these causal processes. That is, the HPC view yields a massive underdetermination problem: Is cancer one kind, or many? And, if many, how many?.

The properties that make cancer "cancer" are not reducible to, or explainable wholly in virtue of, features of cancer cells. Here's an analogy that may make this point more vivid. Cancer is very like aging. In fact, cancer is a disease of aging. Aging is a complex process that involves the breakdown of various physiological mechanisms at a variety of temporal and spatial scales. For each species there is a typical series of ways in which aging progresses. Some of these ways are affected by and associated with changes to genes and chromosomes, but many are not. Shortened telomeres are, for instance, one cause of aging in humans, but by no means the only one (Martin-Ruiz et al., 2006). Assuming for the sake of argument that aging is a natural kind of process, are the particular mutations and genomic changes to cells associated with aging "essential" conditions of aging? If we were to read a paper by cellular or molecular biologists on the hallmarks of aging, chances are we would conclude yes, for such biologists identify aging with aging *cells*. But chances are that a paper by gerontologists would point to a very different set of causes, mechanisms, and pathways associated with aging, at very different temporal and spatial scales. Gerontologists would certainly agree with cell and molecular biologists; their views are consistent, but it does seem odd to insist that the latter have got the "mechanisms" of aging right, but the former do not. This is not to say that such choice of mechanism is arbitrary; it is a matter of what it is that one is hoping to explain or which outcomes are of greatest concern. Gerontologists identify causes, mechanisms, and pathways, no less than molecular biologists do. They simply focus on a different temporal and spatial scale than geneticists.

In summary, the HPC account suffers from three difficulties. First, there is a challenge in establishing that cancer cells are a natural kind. Second, it is far from clear that caretaker mutations are "the" mechanisms for cancer. Indeed, it is not clear whether we ought to call the mechanisms "for" cancer "homeostatic," given that (most of the time, thank goodness) the countervailing mechanisms that prevent cancer more often prevail. Third, it is not clear that even were we to establish what makes cancer cells distinctive as a kind, we could thereby determine what makes cancer "cancer." More generally, what this discussion suggests is that the HPC view seems to shift the burden of demarcating kinds to a much more difficult set of problems: those of demarcating "mechanisms" and determining when a mechanism or class of such mechanisms is "homeostatic enough" to designate a process a natural kind of process. The questions the HPC view raises are several: Exactly how tightly need such properties cluster for classes of entities or processes to count as natural kinds? How heterogeneous might such a kind be before it ceases to be a kind? How are we to decide which mechanisms pick out the correct classification of natural kinds?

There are two ways we can go at this point. One is to give up on the requirement that cluster kinds have shared homeostatic mechanisms. This is the strategy that has been advanced by Slater (2015) in his account of "natural kindness." Another is to focus on a distinct account of kinds, "functional kinds." I turn first to the latter.

1.5. The End of Diseases?

In his provocatively titled "End of Diseases" (2007), Lange argued that while disease categories have functioned historically as medical kinds, in the future molecular medicine "is on its way to rendering diseases obsolete as medical natural kinds." Lange's claim is that advances in molecular medicine suggest that disease categories traditionally understood will simply become obsolete. What has traditionally unified disease kinds is their similar disruptions of functions, or "a disease is a natural kind of incapacity that features in interesting function-analytic explanations of other unhealthful incapacities" (Lange, 2007, 266). In the case of cancer, the incapacity in question, according to Lange, is disruption of "typical patterns of cell birth and death." Lange thinks that cancers (indeed, many diseases), once "well defined at the molecular level," will be found to be "multiply realized," and so:

> Medical explanations of a token illness's manifestations, predictions of its course, and choices among possible therapeutic strategies will increasingly be based not on the disease category to which the token belongs, but rather on the token's specific molecular subtype. A subtype will not be a natural kind of incapacity. Therefore, it will not be a distinct disease. Furthermore, many subtypes will contain only a single token illness. Predictions in medicine will be made not by drawing upon our past experience with other cases of the same subtype, but by inferences from chemical laws and the patient's biochemical state. (Lange, 2007, 283)

Let us unpack this argument in some detail:

1. A disease is a natural kind of incapacity.
2. Advances in molecular medicine seem to suggest that, in the future, cancers will be diagnosed and treated on the basis of distinctive features of their

individual, or "token," molecular subtype rather than their type, traditionally understood (e.g., lung cancer, prostate cancer).

3. Such explanations will involve appeal not to lawful regularities about medical kinds, but to chemical laws and initial conditions (including a patient's biochemistry).

4. Molecular subtypes are not natural kinds of incapacities.

5. Molecular subtypes hail the "end of diseases" as natural kinds.

I have three main concerns with Lange's argument.[19] On Lange's account of disease kinds, we face an analogue of the underdetermination problem we faced with the HPC account. We can give more or less fine-grained characterizations of disruption to function in cancer, but nature does not tell us which grain is "the" natural one.[20] Much like the problem of characterizing the "mechanisms" for cancer, how we draw the boundaries around "functions," and thus which classification of cancer types and subtypes we might arrive at, is underdetermined by the phenomena. We must make choices about what outcome and which scale of analysis—temporal and spatial—we ought to focus our attention upon. Nature cannot make these choices for us. In Chapter 3, this same problem of scale-relativity poses some challenges for the biostatistical theory of disease, which relies on an account of disease as dysfunction.

Nevertheless, the hope that Lange appeals to is one shared by many cancer researchers: using "molecular subtyping" we will (eventually) be able to predict the course of disease and likely responses to treatment on the basis of "chemical laws" and initial conditions (which he describes as the "patient's biochemistry"). I am more skeptical of this eventuality than Lange. While there have been some successes using so-called basket trials for some rare cancers that share unique molecular signatures (Hyman et al., 2015), clinicians are discovering that a drug designed to intervene in a specific pathway in one cancer (e.g., melanoma) may be successful, but the same drug, when used to intervene in that same pathway in another cancer (e.g., pancreatic cancer), may be unsuccessful. Context—not only where a cancer occurs or what the tissue of origin is, but also a suite of other biological factors—matters for the purposes of successful prognosis and treatment.

---

19. To be clear, my aim here is not primarily to attack Lange, with whose views I am broadly sympathetic, but to use this discussion to uncover the challenges facing appeal to functional-analytic explanations as a way of resolving problems of classification in disease.

20. I will not burden the reader with a list of functions disrupted in cancer, among which one might include anything as vague as "disruption of the cell cycle," or "de-differentiation," or something as precise as the role of "UCN-01 (7-hydroxystaurosporine), a protein kinase C inhibitor that may block G_2 checkpoint regulation" (Wang et al., 1996). To get a sense of the challenges involved in disambiguating this problem of which functions are disrupted in cancer and how we demarcate them, I recommend Bertolaso's (2009, 2016) excellent discussions.

Getting drugs to target requires knowledge of unique features of the tissue micro-environment, features that are not reducible to the "biochemistry" of the patient. The tissue organization in which a cancer arises may be enormously important for predicting the course of a disease. Moreover, cancers evolve, or change over time, so their distinctive genetics and the biomarkers they express at a given time may or may not be the ones we need to target later on (Lipinski et al., 2016). The presence of cancer "stem cells" also predicts how likely a cancer is to recur and thus which sorts of treatment regimen are appropriate (Laplane, 2016). *Where, when, and in what particular order and combination* mutations occur, and thus how their downstream products act and interact in the complex tissue environment, make an immense difference to the course of the disease.

In sum, I am skeptical that molecular medicine will completely eliminate the variety of needs (clinical, as well as scientific) that diverse, overlapping classifications of cancers serve. Where does this leave us? We have, so far, arrived at a variety of underdetermination problems facing both the HPC and function-analytic accounts of disease. It is unclear (a) whether there are "naturally" demarcated homeostatic mechanisms for cancer, (b) whether these mechanisms are "homeostatic" enough to count cancer (in general) as a kind, as well as (c) how to pick out the appropriate dysfunctions. Nonetheless, each such account does seem to capture some important aspect of what we wish to know when we classify disease. What, if anything, do such accounts have in common?

One route is to suggest that all natural classifications are ones that figure in causal generalizations, or generalizations that describe "the causal structure of the world." According to Craver (2009), given the pragmatic character of how we demarcate mechanisms (and/or functions) in neuroscience, we ought to opt instead for what he calls the "simple causal view" of natural kinds:

> It is possible to reject [the homeostatic mechanism] and keep the rest as a simple causal theory of natural kinds. According to this view, natural kinds are the kinds appearing in generalizations that correctly describe the causal structure of the world regardless of whether a mechanism explains the clustering of properties definitive of the kind. (579)

The simple causal view may sound intuitively plausible. Indeed, Khalidi (2013) defends a version of this view as one that makes good sense of many of our classification practices in a variety of sciences. The intuition behind the view is that there must be something or other that "grounds" or "supports" our generalizations about natural kinds. What could this be, other than the "causal structure of the world"?

There are two challenges (at least) that this view faces. First, there is (perhaps not surprisingly) some ambiguity in how we ought to apply this criterion for kind status. Which kinds figure in generalizations "that correctly describe the causal structure of the world"? This account may be either very permissive or very restrictive depending upon one's view of causation and what counts as a "correct description" of the causal structure of the world. Consider that consuming bacon may significantly increase one's chances of getting cancer. It may well be a generalization that "correctly describes the causal structure of the world" that people who regularly consume bacon are at risk of cancer. Does it follow that "regular consumers of bacon" are a natural kind? The account seems to trade one underdetermination problem (what it is for something to be a mechanism) for a more difficult family of problems: what it is to figure in generalizations that give a "true causal description" of the world. Indeed, if there is one or more than one "correct" way of carving up the causal structure of the world (see, e.g., Dupré, 1993; Cartwright, 1999), there may well be more than one adequate way to classify kinds.[21] The simple causal view yields a unified taxonomy of kinds only on the assumption that there is *one* correct way of carving up the world's "causal structure." If there are many, then there will be many cross-cutting kinds. Perhaps this is a bullet we can bite. But Craver's intuition seems to be that there must be some principled way of characterizing the causal structure that rules some categories in and others out.[22]

Second, one may well ask, why privilege causal basis as essential to kind status? The intuition seems to be that the causal structure of the world is ultimately what science is aiming to describe. But is describing the causal structure of the world the sine qua non of science? Some philosophers are skeptical; arguably not all laws of nature and not all scientific explanations are causal (see, e.g., Lange, 2016; Huneman, 2010). It seems that we can serve some of the goals of science by identifying kinds that figure in causal generalizations, but perhaps not all. Ereshefsky and Reydon (2015) (as we shall discuss further) argue that scientists seem to be willing to endorse a variety of "non-causal" kinds. For instance, microbiologists' "phylo-phenetic" species concept relies almost

---

21. Thanks to a reviewer for pointing out that "this last question is a substantial one, and the author is right that many sidestep it . . . I recommend . . . expanding it in a way that presses the problem of multiple causal individuations or offers some other novel contribution to the difficulty." This book is essentially my reply. One of the central arguments of the book, in fact, is that there are many disciplinary perspectives from which, and temporal and spatial scales at which, we may identify and characterize causal explanations of cancers. Thus, it is not surprising that different cancer researchers interested in different causal explanations classify cancers differently.

22. Woodward (2003, 2010, 2016) as well as Spirtes et al. (2000) provide a far more sophisticated discussion of causal modeling and causal reasoning than I can provide here, given limitations of space. Reiss (2015b) has also defended a view of causation, evidence, and inference with which I'm broadly sympathetic. See also Illari et al. (2011) for a set of essays that cover the spectrum of modes of reasoning about evidence for causation in the sciences.

entirely on non-causal features. Functional kinds, for instance, of genes, demarcate kinds in terms of their functional role, whether or not they have a common causal basis (though, arguably, serving this functional role is a kind of "fitting into the causal structure of the world" in a very general sense). What they call "heterostatic" kinds are groups that have a number of similar properties but that also have persistent differences within them. These kinds lack a common causal basis, yet they appear to be (more or less stable) categories that enable robust predictions.

These sorts of considerations might lead one to go "minimalist." That is, perhaps kinds are simply collections of entities or processes with more or less "stable" properties, whether or not these are unified by a common set of homeostatic mechanisms or functions. This is Slater's strategy in the "Natural Kindness" account of kinds. Slater (2015) argues that stability, and thus "naturalness" or "kindness," is a matter of degree. According to Slater, we ought to jettison the requirement of a causal basis for a kind, and instead simply take the relative stability of clustering of similar properties, behaviors, or incapacities as the foundation for kind membership. Slater calls this "cliquish" stability: "Properties are clustered in such a way that possession of some of them reliably (if imperfectly) indicates the possession of whole cluster (if not each property in the cluster) at that time. It need not imply that a particular that possesses any of these properties will continue to possess them" (2015, 397). The advantage of Slater's view is that it accommodates relatively ephemeral or dynamic kinds, such as species (and cancer, or evolving populations of cells and cell lineages, or, for that matter, ecological communities). The disadvantage, however, is that it saves the phenomena at the cost of dropping almost all but one prior requirement on natural kinds. It would seem to be so permissive as to leave indeterminate which clusters count as "stable" enough. This line of argument echoes an objection I raised earlier to the HPC view: How stable need these clusters be to count as "kind-like" or "kind" enough?

Slater has an answer to this objection. According to Slater, the stability of a cluster must be "consistent with the probabilistic entailment relationships from sub-clusters to clusters" and "be consistent with the natural laws of some domain and meet the relevant applicability standards" (2015, 400). In other words, a cluster is stable enough to count as a kind if it stands in the proper relationship with other kinds for some domain and its behavior is consistent with the natural laws of some domain. The domain-relativity of laws of nature is a notion Slater borrows from Lange (2005). According to Lange, laws are distinct from accidental generalizations in light of the former's range of invariance under counterfactual suppositions. The trick here, however, is which counterfactual suppositions we take seriously. After all, biological and ecological "laws" (such as there are) are not universal, but hold only given that certain conditions are in place; many such laws are less "stable" or "invariant" (Woodward, 2003) than laws of physics (see also Mitchell, 2000).

What, exactly, does it mean to be "consistent with the probabilistic entailment relations" or "laws in some domain"? According to Lange, maximal invariance is not a property of laws individually, but of sets of laws in a domain. A set of laws is "subnomically stable" when all members remain true under any counterfactual supposition consistent with the set. But which counterfactuals we are willing to "take seriously," and thus whether we consider a group of generalizations lawful, is "domain-relative" (or perhaps discipline-relative). For instance, ecological laws take as a given that the earth will continue to orbit the sun and that temperatures will vary regularly with the seasons. The fact that it is physically possible that such states of affairs could fail to hold would not lead ecologists to abandon their laws. This means, however, that on Slater's account, what we are willing to count as a kind for any science depends upon the antecedent counterfactual conditions we are willing to "take seriously" in that domain or discipline. Satisfying the conditions of adequacy for what may count as a kind for ecology is thus going to be rather different than satisfying these conditions in physics. This leads to the potential for pluralistic and cross-cutting classifications of kinds, a position that Slater sees as an advantage rather than a limitation of his view. I agree; though, of course, granting this means that we end up with a much more permissive, cross-cutting, and pluralistic account of natural kinds than is typically assumed. Indeed, scientists endorse far more "natural" kinds than what many philosophers would grant by the lights of their presuppositions about "naturalness" (see footnote 2).

Slater's view nicely dovetails with Ereshefsky and Reydon's (2015) account of "scientific" kinds. This account depends upon the idea of a "classificatory program" (Ereshefsky, 2001). According to Ereshefsky, a classificatory program is the part of a scientific discipline that produces a classification, and it has three components: motivating principles, sorting principles, and classification. Sorting principles sort entities into kinds; motivating principles justify the use of sorting principles; and a classification is whatever is produced by the combination of the two. Table 1.1 lays out the variety of classifications of cancer, based upon our review of cancer classifications in section 1.2, organized by motivating principles, sorting principles, and classification system, drawing upon Ereshefsky's model.

As we can see from the table, different agents with different purposes (surgeons, oncologists, pathologists, epidemiologists) have different epistemic (and practical) interests, which yield distinct sorting principles and distinct classifications of tumor types and subtypes. According to Ereshefsky and Reydon (2015), a scientific classification is sound when its epistemic aims, or the motivating principles, are consistent with and promoted by its sorting principles, the sorting principles are empirically testable, *and* the classification program is progressive rather than "degenerating," a notion they take from Imre Lakatos

Table 1.1 Motivating Principles, Sorting Principles, and Resulting Classification Scheme for Kinds of Cancer

Motivating Principles	Sorting Principles	Classification
Determine where and how much to cut (surgery)	Size, location, TNM	Staging
Predict how quickly a cancer is likely to progress, recur, or (if early-stage), invade (oncology)	Size, number of metastases, mitotic activity, biomarkers, specific biomarkers associated with specific cancer types (tissue, cell type)	Staging, grade, molecular data (proteins expressed, etc.), associated with invasion, recurrence
Predict how a cancer is likely to evolve resistance to chemotherapy (oncology, mathematical modeling of tumor evolution/genetics)	Size, current, or predicted genetic heterogeneity of a tumor, death rate, presence/absence of stem cells/tissue, architecture/presence of fibroblasts	Staging, grade, tumor sampling for heterogeneity, biomarkers indicating stemness and/or tumor heterogeneity
Predict how a cancer is likely to respond to specific, targeted treatment (pharmacy, oncology)	Biomarkers, tissue type (presence of fibroblasts, etc.), networks affected	Biomarkers: e.g., ER-positive, HER-2neu, mutations such as BRAF or EGFR
Determine remote cause of cancer/likely chance of further cancers in other locations (epidemiology, cancer genetics)	Presence of germ-line mutation/patient history	Familial cancer syndromes/patient history (smoking, infection, etc.)
Identify proximate causes of cancer	Cell and molecular biology, cancer genomics/proteomics/transcriptomics, network theory, systems theory	TCGA, cancer subtypes based on shared genetics, etc.

Source: Drawn from Reydon and Ereshefsky's (2015) concept of a classificatory program.

(1980). Lakatos argued that a scientific research program is progressive when it makes novel predictions and, in particular, predicts facts that other theories do not. Likewise, Ereshefsky and Reydon argue, "A classificatory program is progressive if it provides principles that produce additional classifications or extend existing classifications (relative to competing classificatory programs) and those classifications are empirically successful" (982). They contend that many classificatory programs in the sciences meet all these conditions. That they fall short of more stringent requirements on natural kind classifications proposed by competing accounts, in particular the HPC view, is no reason to reject them as legitimate "scientific" kinds.

I am broadly sympathetic with this view; however, there are several concerns one might have. First, perhaps, how are we to assess whether a classificatory program is "progressive"? It is certainly the case that false theories can at least appear to be empirically successful relative to rival theories, sometimes for quite a long time (see, e.g., Chang, 2010; Stanford, 2010). Second, and relatedly, Ereshefsky and Reydon are defending an account of "scientific" kinds, but are these "natural" kinds? That is, have they impermissibly shifted from a metaphysical question to an epistemic question? Successful scientific research may, after all, not give us access to truly natural classification. To some extent, Ereshefsky and Reydon bite this bullet; their audience is philosophers of science, not metaphysicians. They explain that their goal is to provide "an account of kinds that better captures the variety of classificatory practices found in science" (970). One might worry that whatever classificatory practices scientists use, these may or may not track "natural" kinds, but may be in service of scientists' particular (and potentially idiosyncratic) interests or purposes. Franklin-Hall (2015) raises exactly this objection to what she calls "the simple epistemic view" of natural kinds. This is the view that "the natural kinds are groups corresponding to categories that best serve our epistemic aims, most centrally our project to represent aspects of the universe accurately, as well as to make correct predictions and offer up explanations about the phenomena within it that interest us" (938). She argues that this falls short as an account of natural kinds, because our epistemic interests may or may not lead us to arrive at a genuinely "natural" classification. She calls this the "coordination problem":

Why have categories determined in part by us coordinated with natural kinds determined in no way by us? Without an account of this, our successful identification of the natural kinds via the scientific categories will appear to be a remarkable instance of cosmic good luck. In particular, we were lucky that we developed interests that lined up so well with the mind-independent natural kinds. (934)

How seriously ought we to take the coordination problem? And how may we overcome this problem? Franklin-Hall (2015) suggests the following solution: a "more objective" classification is one that does not merely serve one group of scientists' interests, but would in principle serve many inquirers' interests. She calls this the "categorical bottleneck" account: "the natural kinds correspond to those categories that are metaphorical 'bottlenecks' in the following sense: they reflect *the categories that both ourselves and a large array of scientific inquirers with epistemic aims and cognitive capacities differing from our own would sanction in common, thereby converging on a single set of categories and kinds from multiple, distinct starting positions or points-of-view*" (940). This view nicely captures what, it seems, most philosophers mean by or hope for in an account of natural kinds. The only problem is, how are we to know what many different inquirers' interests would converge upon, apart, of course, from what they *do* happen to converge upon, over the long run? And why suppose that inquirers with different interests would indeed converge on *one* true classification?

The question Slater's, Franklin-Hall's, and Ereshefsky and Reydon's accounts raise is the following. Why ought we to suppose that, however empirically successful our classification, it is the one that tracks "nature's joints"? This is, in essence, Goodman's problem (or the problem that most folks assume Goodman left us with). Recall Goodman's discussion of the predicate "grue":

> [L]et me introduce another predicate less familiar than "green". It is the predicate "grue" and it applies to all things examined before [time] t just in case they are green but to other things just in case they are blue. Then at time t we have, for each evidence statement asserting that a given emerald is green, a parallel evidence statement asserting that emerald is grue. And the statements that emerald a is grue, that emerald b is grue and so on, will each confirm the general hypothesis that all emeralds are grue. Thus according to our definition, the prediction that all emeralds subsequently examined will be green and the prediction that all will be grue are alike confirmed by evidence statements describing the same observations. But if an emerald subsequently examined is grue, it is blue and hence not green. Thus although we are well aware which of the two incompatible predictions is genuinely confirmed, they are equally well confirmed according to our current definition. (1955, 74–75)

Goodman calls his problem the "new riddle of induction." He takes his riddle to provide a challenge to our ordinary assumptions about warranted inference. In particular, the "grue" puzzle is meant to show that "entrenchment is all there is to projectability." The "projectability" of a predicate is just the ability to make inductions or projections into the future about the stability or persistence of the property to which that predicate refers. What Goodman's puzzle suggests is that what enables us to make predictions about green objects is no different from, and entirely compatible with, what might enable us to make predictions about grue objects. Hacking puts this best: "What is the criterion of projectability? Goodman's answer was radical. A predicate is entrenched if it is in use and has been so for some time. The greater the history of usage in successful induction, the greater the degree of earned entrenchment. There is no deeper or more philosophically instructive criterion than that. Most readers have balked at this extreme nominalism. That cannot be the end of the matter!"(Hacking, 1993; cited in Elgin, 1997, 213). Nominalism is a rejection of universals; in this context, this means that Goodman is saying there is nothing in the world (apart from habits of entrenched use) that "grounds" our inductive inferences. Our classifications of things are not identifying universals, properties, or laws that cut nature "at its joints," but are simply ways of cutting up the world that pick out some classes similar in some respects and not others. Beginning with Quine's discussion of Goodman, many philosophers have thought that, as Hacking says, "that cannot be the end of the matter!" Goodman's discussion is taken to vindicate the need for natural kind talk. In other words, what the history of the literature on natural kinds seems to suppose is that there must be such distinctive features of the "natural" or, alternatively, something that all "natural kinds" share that non-natural kinds like "emeroses" or "grue" things do not.

How can cancer classifications be brought to bear on this discussion? We can, of course, find any similarities or dissimilarities we like among cancers. All cancers are (in some sense) alike in some ways and different in others. But this should not be taken as a counsel of despair—suggesting that any form of classification is ultimately arbitrary. A classificatory scheme can be conventional, but not arbitrary. Goodman's famous "grue" challenge was taken to suggest that philosophers needed to find some way to demarcate artificial from natural kinds or determine (in general) which properties are projectable. His answer was "the entrenched" properties. But this leaves us with a puzzle: which properties are entrenched in a "gruish" way and which ones are not? How can we know?

The answer I wish to suggest is as follows. In biology, (and perhaps also in many other sciences), there is no hard and fast line between the gruish predicates and the non-gruish ones. The relative stability of our classificatory categories is a matter of degree, and the tolerance of gruish predicates we are willing to (and ought to) subject ourselves to depends importantly on what we might want to use our classifications for. The "stability" of our categories, and indeed our lawlike generalizations, is a matter of degree. On a suitably permissive view of laws or "lawlike" generalizations, there are simply more or less stable or invariant generalizations in biology (Mitchell, 2000; Lange, 2009, 2013). Such generalizations pick out properties that are more or less projectable (or kinds whose behaviors and dynamics are more or less predictable). Which causal regularities or generalizations we regard as "stable enough" must depend on the temporal and spatial scale of prediction and explanation for some scientific domain. Kinds that figure in these generalizations are the kinds we countenance as "natural" enough. Does this mean we ought to give up on the "realism" of our classification? No, but it may well require that we give up on a view of kinds that is so restrictive as to rule out all but physical or chemical kinds (and perhaps even those; see, e.g., Needham, 2011).

Acknowledging the diverse, cross-cutting classifications we find in scientific practice does not require that we endorse anti-realism about kinds (Khalidi, 1998, 2013; Chakravartty, 2011). Saying why requires attention to the context-sensitivity of property instantiation of higher-level kinds. Higher-level kinds—the kinds of the "special sciences"—are often heterogeneous and unstable in property realization. That is to say, their properties are often realized or activated in a context-sensitive way. And the lawful regularities that describe their properties will mark this context-sensitivity by characterizing these lawful regularities as "ceteris paribus." Lawlike generalizations in many parts of the sciences are ceteris paribus, where what "ceteris paribus" means is often not specified, but could mean either that certain conditions are held fixed or assumed to be constant, when they are not (Cartwright, 1983), or that laws are only "invariant under every counterfactual supposition *of interest to the science* and consistent with the set" (Lange, 2002, 416). In other words, what scientists are willing to consider laws, or lawful enough, is domain-relative.

As we have seen in the context of cancer, the properties and lawlike generalizations about cancer kinds are often highly "unstable" or context-dependent in their realization. Possession of a gene, gene family, or genetic network does not (or not always) determine how a cancer is likely to behave; likewise, cancers that all come from a specific cell or tissue type or organ do not all behave the same way. Scientists are willing to tolerate this kind of uncertainty; indeed, what's the alternative? Scientists are interested in how (all else equal) cancers with similar features behave, and how and why those behaviors come about. They thus may classify kinds of cancer in light of a variety of shared features (e.g., cell and tissue type, and/or genetics, and/or tissue architecture), implicitly or explicitly "hold fixed" some conditions known to vary, and arrive at more or less stable generalizations about cancers with these features. Cancer scientists know (darn well) that cancer kinds can misbehave, or fail to behave as we predict. But this means, effectively, that the generalizations we treat as robust, and the associated classifications of natural kinds in light of their properties, behaviors, or functions, are only "relatively" stable, or stable enough, given our motivating principles. Moreover, this means that the same cancer could belong to one, or several different categories, depending upon what our classificatory aims are. In other words, kind membership is only "partial" in the sense that a cancer will behave like other cancers of its "kind" in many respects, but not in all.

Insofar as domains of inquiry are generated by us and our investigative practices, it would seem to follow that such classifications are shaped at least in part by our interests. These interests help us identify regular behaviors of types and subtypes of cancers. Scientists know that the exhibition of properties broadly shared by such cancers are context-sensitive in their realization, and multiply realizable. It seems problematic that we should be forced to the odd view that such kinds, properties, and regularities are not "real" or "natural" simply because the world does not cooperate by giving us uniformity, much less "necessity." While the paradigmatic cases philosophers have tended to focus their attention on have (or seem to have) essences, it is not clear that most examples of the kinds of kinds scientists investigate do (if any) (see, e.g., Needham, 2011).

What properties are relevant to one's cancer classification depends to an important degree on the subcategory or subtype we happen to be classifying, as well as one's motivating principles. By way of analogy, consider species classification and, in particular, the lineage-specific knowledge that experts in the classification of various subgroups require. An expert on mosses will know which sorts of properties are of greatest relevance to classifying mosses; an expert on tropical birds will, likewise, attend to very different facets and features of tropical bird diversity. What is relevant to explaining and describing diversity in different taxa—with different phenotypes, genotypes, developmental trajectories, life histories, ecologies, and so on—is rather different. This is no less true for the classification of various types and subtypes of cancers. Just as species have distinctive genealogies, ecologies, behaviors, ontogenies, and genetics, so too cancers arising from different remote causes, in different organs and tissue types, require attention to these detailed differences. These differences matter when we are giving explanations and making predictions (and retrodictions) about the variety of things we care about: how cancer is caused and how it's likely to behave. The category of "relevant" similarities is far from uniform, but varies with our domain of inquiry.

Cancers' classifications are not all grounded in the same way, and what sorts of behaviors are typical of these processes vary with circumstance. As Chakravartty (2011) argues, a realist might take conflicting classifications to be genuinely tracking kinds, where different classifications are simply different manifestations of the fundamental kinds' typical behavior: "The fact that dispositions are often manifested differently, depending on the circumstances, furnishes the key to an alternative response to the challenge presented by variable descriptions of the fine-grained metaphysical natures of some scientific entities. One and the same entity may behave significantly differently in different circumstances, even when the properties associated with it are preserved from one circumstance to another. In such cases, on the dispositional view, different behaviors are simply different manifestations of one and the same property (or properties)" (176).

It seems very difficult to do scientifically informed metaphysics that does not—in some way or another—grant that what we countenance as dispositions have different manifestations in different contexts. That property-instantiation is context-dependent seems a basic fact of the natural world; indeed, knowledge of how and why contexts vary in ways that enable reliable predictions about the same kinds of entity is what makes modern medicine and the study of complex heterogeneous diseases like cancer possible. That this is a violation of an essentialist view of kinds seems reason to reject an essentialist view, not necessarily reason to resist the classifications of cancer(s) as natural.

1.6. Conclusion

The question "Is cancer one kind or many?" presupposes that there must be one way in which cancer kinds can be classified in virtue of their "natural" properties or features. But there are many "natural" features of relevance to the classification of cancer, and they do not yield a unified classificatory scheme. Perhaps ironically, to be a good naturalist *and realist* about natural kinds requires granting that the "natural kind" category is itself not very natural. There must be some empirical warrant for categorizing kinds in the ways we do. But the case of cancer suggests that which empirical facts and which features or properties are appropriate for classifying kinds in one domain may or may not be appropriate for all contexts or other domains. It is surely true that cancer(s) have shared features, in virtue of (*very broadly speaking*) shared properties, causes, and mechanisms. But it is also true that there are many ways of picking out kinds of cancer. Depending on the scale of analysis and the type of feature one chooses, cancers cluster in different ways. Indeed, a similar point has been made regarding classifications of biodiversity (see, e.g. Sarkar, 2005).

The very idea that there could be only one true classification—let alone that this classification ought to be dictated by genetic features of cancer cells—is a metaphysical thesis that cancer scientists would do well to disavow, for it leads to ignoring the rule of total evidence. We ignore this rule at our peril; doing so may lead to myopia about not only cancer classification, but also diagnosis and prognosis and, worse still, to narrowing our investigations into cancers' causes and possible avenues for treatment. Genetic and molecular data is useful for some purposes, but as the exclusive guide to classification it is impoverished.

Stanford (2017) and Woodward (forthcoming), for instance, recommend "quietism about metaphysics": "the pragmatist philosopher should just decline to do metaphysics or make ontological claims or take a stand on the issues about grounding, truth conditions and so on, that dominate contemporary metaphysical discussion" (Woodward, forthcoming). Woodward notes that metaphysical discussion "as currently conducted" abstracts from the "human limitations" and "methodological and interpretive issues that are the province of the pragmatic philosopher of science." I am very sympathetic to the pragmatist's intolerance for metaphysical debates that trade on competing intuitions with little or no connection to the world. But surely some metaphysical debates are worth our attention? Which sorts of metaphysical debates are worth our attention? Is the matter of what count as kinds, as Hacking (1991, 2007) claims, simply a failed research program?

Some parts of this program, in my view, ought to be set aside. It seems clear—if the work of Hacking, Khalidi, Slater, Dupré, Kitcher, Griffiths, Ereshefsky, and others is any indication—that the hope for a universal set of necessary and sufficient conditions for kind membership that covers all natural classifications is defunct. But we may well endorse a much more pluralistic account which grants that different ways of clustering kinds—some that pick out shared mechanisms, some that pick out shared dysfunction, and others that may have no common causal basis—are equally successful at identifying common kinds of entity and process.

Some may worry that this account is too permissive. If all that is necessary for counting some class of entities as a kind is that it seems to figure in scientific generalizations, almost anything might count! After all, we can make predictions about the behavior of philosophers, bachelors, and pencils on the basis of properties that cluster. I agree: the semantics of natural kind terms is not unique to natural kinds; these same semantics are shared with other terms. Natural kind terms are supposed to be special in terms of what they refer to; they are not semantically special. This straightforward observation, however, has mistakenly led to a search for some unified way to "ground" the "naturalness" of natural kind terms. For instance, natural kinds are sometimes said to be distinctive in virtue of their shared "causes" or "intrinsic" or "internal" nature. To some extent, (some) cancers progressively acquire an "internal" nature or come to constitute a set of causal interactions that lead to (relatively) stable outcomes: metastasis and death. The problem is that the homeostasis of this process is a matter of degree; when and why we might wish to say that this process is somehow integrated or "stable enough" to be counted as a natural kind of process is a pragmatic matter. We must decide how tolerant we are of violation of expectations—and cancer is, if anything, a highly volatile example of how our expectations can be violated. The progression to metastatic disease is subject to contingent events at a variety of temporal and spatial scales, which can lead to suddenly aggressive disease or to lifelong "indolent" tumors, which never progress. The very idea of homeostasis in light of "stable" causes or "internal nature" in such cases breaks down; indeed, this is not unique to cancer, but seems to be a problem for many paradigmatic cases of biological kinds.

Thus, part of the burden of this chapter has been to challenge various philosophers' attempts to offer monistic criteria for grounding the "naturalness" of kinds. It is tempting to say that the "causal structure of the world" is what glues the kinds together, but without specifying what this means, this is just another way of saying that something or other explains that kinds are more or less reliably co-occurring. Different disciplines (for good reasons) offer very different stories about what is doing the "gluey" work or how stable or homologous kinds need to be. With Goodman (1955), I am skeptical that we may make a distinction other than (perhaps) one of *degree* of stability between laws and accidental generalizations.

Classifications in science are founded in inductive methods. Generalizations about natural classes based on shared properties, behaviors, or patterns in the phenomena in scientific practice are almost always ceteris paribus, even in the physical sciences (Lange, 2002). The kinds of generalizations of cancer types and subtypes that hold are no exception; they have a range of stability or invariance (cf. Woodward, 2003). Philosophers who imagine that science will provide a way to support the view that essential properties ground kind membership or that kinds figure in laws of nature (traditionally understood, as both universal and necessary) run aground of the unfortunate fact that such laws seem few and far between in the natural sciences. There is a degree of counterfactual stability of generalizations that we treat as lawful in the sciences, and the counterfactuals we take seriously are often domain-relative (Lange, 2009). Cancer science is no exception. The result, it turns out, is a multimodal and cross-cutting family of classificatory schemes.

FROM DISEASE TO RISK

2.1 Introduction

In the early 1990s, there was a spike in the incidence of prostate cancer in the United States, rising to a peak of 237 per 100,000 per year among US men in 1992 (Siegel et al., 2016). This was due to the widespread use of the PSA, or prostate-specific antigen, test in the 1980s.[1] To some extent, the spike in incidence was expected; a new screening regimen is bound to reveal a reservoir of previously undetected cancers. Indeed, this is one of the rationales for screening. Presumably, if we can catch more cancers early, we will prevent more early deaths from the disease. Since this peak in the 1990s, there has been a decline in incidence (Herget et al., 2016). However, the mortality from prostate cancer has not fallen off proportionately, given the marked improvement in early diagnosis. Why not?

---

1. This test is used to assess levels of prostate-specific antigen, a protein produced by prostate cells. PSA levels higher than 10.0 ng/ml or more are correlated with a greater than 50% chance of finding cancer with a prostate biopsy (SWOP, 2015). The PSA test is highly sensitive. Epidemiologists make an important distinction between sensitivity and specificity. Sensitivity is the probability of testing positive if disease is present, or $a/(a+c)$, where a is a true positive, c is a false negative. Highly *sensitive* tests detect all or most the disease, at the expense of lots of false positives. This is good if the cost of missing a disease is high and the cost of false positives is low. *Specificity*, in contrast, is the probability of testing negative if disease is absent, or more precisely: $d/(b+d)$, where d is the chance of a true negative, and b is the chance of a false positive. A highly *specific* test has few false positives but may miss some of the disease. This is good if you are screening for an uncommon condition where the cost of false positives overwhelms the advantage of finding disease. In general, cancer screening requires *specificity* over *sensitivity*.

	Test positive	Test negative
Tumor	A True positive	C False negative
No tumor	B False positive	D True negative

This is an empirical question, and it is (not surprisingly) a contested one. Some suggest that there simply has not been sufficient follow-up time to observe what will ultimately be a compensatory drop in mortality. However, it is more likely that some proportion of these cases were overdiagnosed and overtreated. "Overdiagnosis" is the identification of disease that would never have led to symptoms in the lifetime of the patient. Overdiagnosis can occur when someone is diagnosed with an indolent or slow-growing disease or is diagnosed at such an advanced age that the person is unlikely to suffer harm from the disease diagnosed. Overtreatment is treatment that would not have prevented clinical symptoms or death. There is no question that prostate cancers have variable natural histories. Some progress quickly to metastasis, while others are slow-growing or even indolent. Many men die "with" prostate cancer but not "of" prostate cancer. The average eighty-year-old man in the United States has about a 50% chance of having "pre-cancerous" lesions (hyperplastic, dysplastic, or even neoplastic lesions) in his prostate (Welch et al., 2010). Many of these lesions will never present symptomatically in these men's lifetimes.

On the basis of two recent clinical trials, one in the United States and a second in Europe, the number of men overdiagnosed for prostate cancer was estimated to be as high as 40–50% (Liong, 2012, e45803). On the one hand, this may seem to be significant and harmful. The quality of life for men treated for prostate cancer may be poor, because removal of the prostate may cause incontinence or impotence. Treatments such as "chemical castration" can have other side effects, including weight gain and loss of sexual function. On the other hand, it may seem that, all things considered, treating these men was not a harm. After all, we could not (and in many cases, still cannot) determine which cases were overdiagnosed, and so it was better, all things considered, to be safe than sorry. However, such cases do seem to involve a significant trade-off, one that entails a balance of personal preferences and tolerance of risk.

Is this problem unique to prostate cancer? The balance of evidence suggests that there may well be other cancers that have likewise been overdiagnosed and overtreated, including thyroid, breast, and lung and skin cancers (see, e.g., Esserman et al., 2014). The United States Preventive Services Task Force (USPSTF, 2016) notes that there has been an almost 50% increase in the rate of diagnosis of breast cancer during the era of mammography screening. While they argue that "it is not possible to know with certainty what proportion of that increase is due to overdiagnosis and what proportion reflects other reasons for a rising incidence," they also argue that it is possible that as many as "1 in 3 women diagnosed with breast cancer today is being treated for cancer that would never have been discovered or caused her health problems in the absence of screening. The best estimates (cf. Marmot et al., 2013) . . . suggest that 1 in 5 women diagnosed with breast cancer over approximately 10 years will be overdiagnosed" (USPSTF, 2016). To be sure, these numbers are contested. One review (Puliti et al., 2012) identified twenty different estimates of overdiagnosis in breast cancer based on long-term follow-up studies of mammography screening trials, ranging from less than 10% to as high as 60%.

Concerns about the extent of overdiagnosis and overtreatment have led some epidemiologists to question whether we should screen as often as we do, particularly for prostate and breast cancer (see, e.g., Gøtzsche and Jorgerson, 2013). However, national and international organizations' recommended changes in screening have been met with a firestorm of opposition. The American Cancer Society and professional organizations like the American College of Radiology (ACR) and the Society of Breast Imaging (SBI), for instance, continue to recommend that women get yearly mammograms starting at age 40. This is despite the fact that the USPSTF concluded that the evidence of benefit, in terms of lives saved, was small compared with the cost overall, including unnecessary screening and biopsies, worrying patients, and overdiagnosing and overtreating, apparently, some not insignificant percentage of patients.

Moreover, recommendations have been issued for relabeling previously identified carcinomas as instead "indolent" or non-progressive forms of disease. A team associated with the National Institutes of Health (NIH) has recommended that ductal carcinoma in situ (DCIS) be "downgraded" from "carcinoma" to IDLE (idiopathic lesions of epithelial origin) to signify that DCIS is not the same disease as invasive breast cancer and does not warrant aggressive treatment (Esserman et al., 2014). The latest cancer to come under scrutiny is thyroid cancer; the encapsulated follicular variant of papillary thyroid carcinoma has been relabeled as "non-invasive follicular tumor with nuclear features of papillary thyroid cancer" (or "NIFT-P" PMID: 27078145). The concern behind these changes in nomenclature is that many patients may have been undergoing unnecessary treatment for "cancer," when such states may or may not have progressed to metastatic disease.

Stepping back, there appear to be two kinds of uncertainty at stake in these debates, "prospective" and "retrospective." First, there is genuine uncertainty among pathologists about whether some states ought to be considered "cancer." Some "borderline" melanocytic tumors, ovarian tumors, mucinous breast lesions, and soft-tissue tumors are simply "tumors for which it is difficult and sometimes impossible, even for expert pathologists, to accurately predict biologic behavior on the basis of the pathologic features" (Scolyer et al., 2010). What ought one do in such cases? Some authors remarked that a pathologist might simply choose to err on the side of caution:

> When there is genuine uncertainty about whether a tumor is benign or malignant, the most comfortable option may be to call it malignant. This

avoids the possibility of subsequent legal action against the pathologist if the patient develops recurrent or metastatic malignancy that has serious or even fatal consequences. If a nonmalignant or equivocal pathologic diagnosis is provided and the clinician does not undertake treatment appropriate for a malignant condition, the pathologist might be held at law to have been responsible for the patient's bad outcome. (Scolyer et al., 2010, 1771)

Such cases are "prospectively" uncertain.

Second, "retrospective" uncertainty concerns the effectiveness of cancer screening, or the extent to which screening has (on balance) benefited patients. In these cases, while the epidemiological data seems to suggest that we may be overdiagnosing and thus overtreating some proportion of cases, the exact proportion is unknown. Here, the uncertainty is founded not on a lack of knowledge about pathology (though the failure may be traced to unwarranted judgments about pathology), but on the indeterminacy of retrospective studies, which attempt to extrapolate rates of overdiagnosis from either retrospective examination of the relative incidence and mortality from cancer in mammography versus control groups, long-term follow-up data on populations that underwent clinical trials, or autopsy studies conducted on individuals who die from non-cancer-related causes (Welch and Black, 2010).

Both cases involve what philosophers of science call "underdetermination." Determining whether a particular case counts as cancer or, more precisely, is likely to eventuate in invasive disease and harm is not always straightforward. Decisions about what criteria suffice to count some dysplastic growth as cancer are, though empirically supported, always underdetermined by the evidence to hand and so in part involve conventional, value-laden decisions. That these decisions are *conventional* ones does not mean that they are wholly *arbitrary*, however. Consider one case in point: the World Health Organization (WHO) recently revised its standards for diagnosing acute leukemia. The cut-off for acute leukemia was 30% blast cells (immature white blood cells) until 2000, when the WHO decided to change the cutoff to 20%. This was based on the fact that most cases (up to 80%) between 20 and 30% develop or progress quickly (M. Salama, personal communication, 2012). So what is called "acute leukemia"[2] is determined in part in light of empirical evidence and in part by experts' judgment of prospective risk. However, this is a collective or conventional decision made by an expert body,

---

2. Cases with less than 30% blasts would formerly not have been considered to be some other type of leukemia but more likely a myelodysplastic syndrome. (Thanks to Dr. Ian Hageman, Pathology, Washington University in St. Louis for this update.)

not a simple or direct extrapolation from the evidence, and so could well be revised in light of new evidence, or in light of new agreed upon standards for which risks are more or less acceptable.

Likewise, assessments of the effectiveness of screening and rates of overdiagnosis are contested exactly because the evidence underdetermines experts' conclusions. So, precautionary values come into play. In order to assess screening's effectiveness in reducing mortality from a particular type of cancer, one needs an estimate of background or baseline incidence and mortality from a particular cancer in a given population (a group of individuals with a particular age range or sex). That is, one needs to know how many individuals would have gotten cancer and how many of these would have died from cancer without screening. But once screening has already become the standard of care, estimates of baseline incidence and mortality are difficult to arrive at. Various indirect sources of evidence of variable quality are thus appealed to, including historical epidemiological data and long-term follow-up data from the original clinical trials. Which type of data to trust, and how decisive it is, is a contentious matter. For instance, one estimate of baseline incidence is arrived at by subtracting "catch-up" cancers in unscreened groups from the total cancers in the screened group, as measured in the original clinical trials. But problems of underdetermination, and thus inductive risk,[3] in such contexts abound. Indeed, they come into play at several points: in assessing "baseline" risk of cancer, or the underlying rates of cancer incidence and mortality, in evaluating the quality of the clinical trials themselves, as well as the long-term or follow-up data from clinical trials, and in assessing the evidence from autopsy studies. In other words, it is not simply an empirical matter, but also entails evaluative judgments about the quality of research, data gathering, and data assessment that are in play in such cases. Perhaps needless to say, judgments about overdiagnosis are contested exactly because they bear on such matters as how to weigh the value of a life against the costs and harms of unnecessary screening mammography—where these harms range from financial to psychological and physical.

The worry both cases raise is that these values are potentially pernicious. For instance, the pathologist's judgment that an uncertain case ought simply be

---

3. Inductive risk, as defined by Heather Douglas (2000), is "risk of error" in inferring a scientific hypothesis; she argues that non-epistemic values enter into scientific inquiry in cases where risk of error has non-epistemic consequences (i.e., in a scientific inquiry into the effects of dioxin on humans or the environment or, of course, in medicine). One might speak of "epistemic risk" more broadly (following Biddle and Kukla, 2017) as "any risk of epistemic error that arises anywhere during knowledge practices." This could include not only accepting false hypotheses but also making upstream methodological choices, such as characterization or classification of data.

assigned the status of cancer to avoid legal recourse would seem to be a pernicious or poorly motivated choice, one that may well have led to the overdiagnosis problem in the first place. This case may be an instance of a more general phenomena, what some scholars characterize as "risky medicine": the expansion of medicine into the treatment of "disease risk" as opposed to disease (Aronowitz, 2010, 2015). While it is certainly advisable to treat some risky states so as to prevent disease, there is a growing concern that "risky medicine" has gone too far. As historians have documented (see, e.g., Greene, 2007; Aronowitz, 2015), economic and institutional forces appear to have played an important role in the expanded reach of medicine. Where and how to draw the line between "disease" and "disease risk" has become an increasingly contested question. Whether or not there may be overtreatment for a range of conditions—from depression to "pre-hypertension" and "pre-diabetes"—has become an issue of public debate. At the root of these debates are several distinct issues, which can be somewhat difficult to disentangle:

- Are these categories (e.g., "prehypertension," DCIS) genuine diseases or merely disease "risks"? To what extent are such conditions so called because of empirical underdetermination, resolvable by empirical means? To what extent do they reflect an overextension of medicine? That is, what role ought medicine to play in treating risky conditions? How much treatment is "overtreatment"?
- To what extent (and how) are value judgments involved in pathologists' assessments of these states?
- To what extent (and how) are value judgments involved in judgments concerning intervention?
- To what extent are these values pernicious? That is, are these judgments motivated by interest in economic gain (e.g., the pharmaceutical industry's "expanded reach") or perverse incentives (e.g., defensive medicine, or, perhaps a fee-for-service model that reinforces "more is better" in medicine)?

These questions—with a special focus on cancer—will be shape the discussion in this chapter. As the above examples demonstrate, cancers that originate in the same tissue, and even cancers that (at first) may appear to be quite similar pathologically, may be heterogeneous in their rate and character of progression. Disorderly growth in the same tissue or organ can progress to metastasis at variable rates or, alternatively, simply not progress. A diagnosis of cancer (especially in its very early stages) is thus not (or not only) a direct and unproblematic application of empirically founded criteria, but an assessment of the risk a given physical state poses to a particular patient. Any assessment of risk involves some uncertainty. Where to draw the line in early-stage diagnosis is not strictly a matter of determining whether a given state in fact compromises organ or organism *function*, but whether it is *likely* to do so, and whether such risk warrants action.

Some naturalists argue that any line-drawing decision will be "merely academic" (Boorse, 1977), for functional disruption is a matter of degree. But dismissing such matters as merely academic is a mistake, in my view. In fact, addressing them squarely is one of the central challenges of medicine as practiced (at least in the developed world) today. At stake in such cases is not (or not only) whether decisions about how to draw the line (e.g., where there is uncertainty about disease course) are *influenced by values*, but also, whether the influence of such values is *pernicious*.

In some sense, the debate over normative versus naturalist approaches to disease is orthogonal to this question; one could argue that they are simply different debates altogether, for the question at issue is not whether we can arrive at a naturalistic account of disease or health, but whether and how to intervene. But this is too quick. How we characterize an instance of disorderly growth as "cancer" is not simply a matter of academic convention, nor is it exclusively to do with intervention. Instead, the role of values in early stage cancer diagnosis, and thus in defining "disease" and demarcating "disease" from mere "disease risk" more generally, raises questions for philosophy and sociology of science about evidential underdetermination, risk assessment, and how to appropriately negotiate the competing bids of authority among specialists in the contested social epistemology of medicine (see, e.g., Solomon, 2015). In other words, defining "disease" is not merely a matter of conceptual analysis.

To be clear, I don't deny that there is a legitimate role for conceptual analysis. Defining "health" may be important in service of distinguishing "health" itself from the "value of health" (cf. Hausman, 2015), for instance. Hausman persuasively argues that conflating the two can lead to confused attempts to design measures of health's value. What generic health measurements track is not "health itself," Hausman argues, but the ways in which we value aspects or features of our health. It is these evaluative judgments that ought to guide health policy and the allocation of healthcare resources. What Hausman's analysis demonstrates is that what matters to us in the context of health policy is not whether some outcome counts as an instance of "health," but how we value that outcome and what work we want our measures of that value to do for us.

So, conceptual analysis is important; and offering up a conceptual analysis of the concept of health, as well as disease, can clarify a debate, e.g., about the aims of health care. However, there is a trade-off involved in any attempt at definition that aims to be both comprehensive and action-guiding. Any analysis of the concept of "health" or "disease" that seeks to provide a set of necessary and sufficient conditions on either is unlikely to both capture the variety of legitimate ways in which this concept is deployed and be normatively guiding. This problem of trade-offs arises again and again when we try to offer up reductive analyses. Whether we are defining "cancer" or "disease" or assessing epistemic conditions on claims about causation, there are trade-offs between giving an account that is comprehensive and one that is specific enough to be useful in service of this or that purpose. The appropriate criteria of application of a concept are often domain-relative and context-specific. Clarity about our goals or the specific context of application is necessary if we hope make to our concepts meaningful and precise. Which methodological norms are appropriate depends on what we want our concept to do for us. This is especially so in the context of biomedicine, where heterogeneity and vague boundaries abound.

Let us turn now to the problem of defining "disease." I consider first what a philosophical analysis of the concepts of "disease" and "health" may be attempting to achieve and, second, how reflection on these debates may or may not resolve the matters that early-stage cancer raises.

2.2. Concerning Definition

One of the central challenges facing competing accounts of "disease" in the philosophical literature is that the goal is unclear. Is the aim to analyze and describe medical practice? Or is the goal to reform and improve upon it? What sort of reform of practice is necessary or, for that matter, possible? If one subscribes to a value-free ideal for science, then removing any taint of values appears to be the goal. If, on the other hand, one disagrees with this ideal, then the goal ought perhaps to be to make transparent when and how values inform scientific practice, not remove or eliminate such values altogether.

The philosophical exchange on disease has been clouded in part by an unclear articulation of and lack of agreement about the goals of conceptual analysis or, perhaps, a more or less Carnapian or Oppenheimian explication.[4] Normative

---

4. "Explication" is a term of art among philosophers, and its sense varies (see, e.g., Schupbach, 2017), but according to Carnap, the goal of explication is to make otherwise inexact concepts precise or to "transform a given more or less inexact concept into an exact one or, rather, to replace the first by the second" (1950, p. 3). According to Carnap, explication does not require a relation of "complete coincidence" between how a term is used in practice (or the explicandum) and the ideal conditions described by the explicatum (the precise concept offered up by the philosopher): "a concept must fulfill the following requirements to be an adequate explicatum for a given explicandum: (1) similarity to the explicandum (2) exactness (3) fruitfulness, and (4) simplicity" (Carnap, 1950, 6). On Carnap's view "close similarity is not required."

Unlike Carnap, Kemeny and Oppenheim (1952) take the aim of explication to be that of "illuminating" rather than replacing the explicatum, and they value descriptive similarity to typical use over other virtues. Schupbach puts this very well: "[I]n Oppenheimian explication, the goal is concept *clarification rather than concept engineering*" (2015, 8). On the Oppenheimian view, one's explication must align fairly well with how a term is understood

theorists hope to provide a critique of what they see as transparently normative judgments in disease attributions. Naturalists have a different kind of reform in mind; their goal is to avoid confusing evaluative judgments with empirical ones. Hybrid or two-stage theorists seek as far as possible to ground disease concepts in empirical science, but take there to be a point at which empirical data underdetermines hypotheses about disease status, and normative judgments thus play a complementary role. Eliminativists contend that, for this very reason (theoretical underdetermination), we simply ought to abandon the project of defining disease in any way that will resolve controversial cases (e.g., obesity). On the eliminativist view, we should simply state the facts in as theory-free a fashion as possible, in order to make transparent when and why value judgments are entering into assessments of various physiological states.

So, each theorist seems to emphasize either more "Carnapian" or "Oppenheimian" goals; some aim to replace our vague concept with a more precise one, and others aim for descriptive adequacy and context-sensitivity. Which of these ways of understanding the goals of definition best characterizes the debate over "disease" and "health"? It's not always clear, and indeed, that may be one reason why this debate has so long gone unresolved. Philosophers of medicine such as Boorse claim that their aim is to offer an account of disease that is in keeping with medical practice; sometimes he describes this goal as "conceptual analysis." However, at other times, Boorse appears to be more concerned with a regulative ideal than with actual practice (Schwartz, 2014). That is, while sometimes Boorse seems to be more Oppenheimian in spirit, at other times he seems to emphasize the Carnapian goal of *replacing* a less precise concept with one that is more precise.

What I seek to do in this chapter is to break through this stalemate, as follows. First, we should grant that there are different and equally legitimate goals that a definition of disease might serve. No definition of disease is likely to meet all such goals simultaneously. General appeals to dysfunction, or the biostatistical theory, are fine for some purposes, such as Hausman's (2015), of distinguishing health from measures of health's value, but they fail to make transparent the role of values in medicine and provide little to no practical guidance with respect to cases of genuine ambiguity about dysfunction or uncertain risk. Cases of early-stage or "borderline" cancers provide a useful way to refocus the debate and uncover problems of underdetermination, both with respect to empirical judgments

---

and used in practice. Of course, all explication is constrained by practice in a way that defining by stipulation alone need not be; if our normative ideals end up counting a significant number of things in or out that are not in ordinary practice, we might say that the explication failed. Nonetheless, how *much of ordinary use or practice we may legitimately jettison* in our explication depends upon whether we are more Carnapian or Oppenheimian in spirit (cf. Schupbach, 2015).

about function and dysfunction and with respect to how values enter into clinical decision-making, whether in the matter of diagnosis or treatment. Let us turn now to the philosophical debate on defining "disease" and "health," with the aim of determining how, if at all, this debate intersects with the questions listed in Section 2.1.

2.3. Naturalism: Disease as Dysfunction

Naturalists claim that we ought to define disease or pathology by appeal to biological theory alone. Boorse (1976, 1977, 1997) is the most frequently cited representative of this view. He defines disease as follows:

1. The *reference class* is a natural class of organisms of uniform functional design; specifically, an age group or a sex of a species.
2. A *normal function* of a part or process within members of a reference class is a statistically typical contribution by it to their individual survival and reproduction.
3. A *disease* is a type of internal state, which is either an impairment of normal functional ability, i.e., a reduction of one or more functional abilities below typical efficiency, or a limitation on functional ability caused by the environment.
4. *Health* is the absence of disease. (1997, 7–8)

Boorse's first condition makes clear that assessments of disease are context-dependent; this condition captures a central feature of the practice of medicine. Clinicians do not describe a postmenopausal woman as diseased because she can no longer have children. The comparison class one chooses when assessing disease status is not that of the species as a whole, but some smaller class of individuals. Function and dysfunction of a part or process ought to be assessed, according to Boorse, relative to the "statistical" mean for one's sex and age. Pathology, on this view, occurs when a part or process performs its function in a way that departs from what is typical for one's reference class.

Central to Boorse's view is that disease is a departure from "species-typical function," which can be understood in terms of *statistical* norms, or ranges, of functional states appropriate for a designated age or sex. Thus, he calls his view a "biostatistical" theory of disease. Such ranges are, in his view, empirical facts to be discovered. That is, there is a fact of the matter about species-typical function for some part or process for individuals of a particular age or sex, and this fact is a product of empirical investigation. This investigation is not simply a matter of sampling; the actual average in a given population (e.g., the average weight of women in the United States) may in fact depart from this ideal. Boorse argues that the goal of physiology is to characterize species-typical functional states, which ought to be based only in part on empirical investigation of the range of variation. Health is "conformity to species design" (1976, 2) in service of survival and reproductive success, and a disease is an "internal state of the organism which . . . interferes with the performance of normal function" (62). Whence, then, does this idea of "normal function" come?

2.4. Problematizing Function

A central issue of contention in the debate over the naturalistic account of disease is exactly this matter of how to define "function" and how to determine when claiming that a state is "dysfunctional" is warranted. Different scientists, in different contexts of inquiry, define "function" in slightly different ways. One may define function in terms of the *causal history* that led to a state (where that causal history included natural selection) or in terms of the current *consequences* of that state for the system in question. Philosophers sometimes distinguish these competing views as *etiological* (causal) versus *consequentialist* views of "function," or "Wright" versus "Cummins" functions, in light of two influential philosophers who defended alternative interpretations of "function" (Wright, 1973; Cummins, 1975).

These competing accounts arrive at different assessments of contentious cases. For instance, on the etiological view, the function of a trait is the activity that natural selection favored. A dysfunction is thus the failure to carry out selected function. So, if in our evolutionary past it was optimal to shore up excess reserves of fat so as to better survive long periods with depleted resources, then "thrifty genes" (presumably genes that somehow or other enable one to better store fat) are functionally optimal (for a discussion of why and how this issue is contested, see, e.g., Gluckman et al., 2009; Genné-Bacon, 2014). In other words, on the etiological view, the function of a trait is contingent on that trait's evolutionary *history*. If a trait is a product of historical accident (e.g., if it were a byproduct of some other feature or trait, or, drifted to fixation, even if it currently enables an organism to better survive or reproduce), then it does not count as a "function" on the etiological view. In contrast, on the consequentialist view, what makes something a function depends exclusively on its current consequences, not its evolutionary history. Thus, given the excess of available resources in our current environment, "thrifty genes" could be considered a "dysfunctional" condition by the lights of a consequentialist.

Boorse is often confused with etiological theorists because of his appeal to the consequences of the disease state for survival and reproductive success. However, it is *current* survival and reproductive success that he is concerned with: "the difference between my view of functional design and Wakefield's is just this: on my view, a functional trait must serve S & R [survival and reproductive success] in the present, while on Wakefield's, it must have served S & R in the past and been selected for that effect" (Boorse, 2014, 557). Species design is, in his words,

> the typical result of evolution ... a trait's becoming established in a species, only rarely showing major variations under individual inheritance and environment. On all but evolutionary time scales, biological designs have a massive constancy vigorously maintained by normalizing selection. It is this short-term constancy on which the theory and practice of medicine rely. (1977, 557)

According to Boorse, traits become established because they allow organisms to achieve optimal survival and reproductive success; and departures from this optimum often coincide with departures from survival and reproductive success in our current environments. Boorse takes "species-typical design" to be whatever is functional with respect to what he calls "benchmark" environments, the most common environments in which humans have lived. In other words, "health" should not be conflated with adaptation to one or another environmental circumstance but with respect to a typical range of environments:

> [A]daptation is not freedom from disease. All sorts of abilities—violin playing, tightrope walking, impersonating a President—may enhance people's ability to live well in their particular environments. But that does not mean that the lack of these abilities would be pathological for them or anyone else. Ordinary medical thought uses no such notion as "pathological for person X in environment E," though "bad for X in E" of course makes sense. The relativity of adaptation to environment, which is its main attraction, is also what makes it unpromising for an analysis of disease. (1977, 549)

For Boorse, function is not the same as adaptation to some specific environment or other. This is why he insists that it is the science of physiology, *not evolutionary biology*, which provides an account of function. The "theoretically normal species design" is whatever best subserves the continued survival and reproductive success of organisms of a given age and sex.

Philosophers have engaged in a long dispute over the appropriate meaning and scope of claims about "function" and "dysfunction" (for an excellent review, see, e.g., Garson, 2016). There are also different "meta-views" about how this issue is to be decided. For instance, "monists" about function argue for one particular account of function. "Pluralists" about function argue that several senses of the term are applicable in different contexts (see, e.g., Godfrey-Smith, 1993). For our purposes, we need not take a stand on this question. Let us, instead, consider the question of how cancer fares on the naturalist account.

On the one hand, it would seem quite obvious that cancer is "dysfunctional"—after all, it is the major cause of death in the United States after heart disease for all people over 50. Cancer would seem to be a clear case of dysfunction. However, in order to answer the question of whether or how cancer is "dysfunctional," we need to specify not only what we mean by "function," but also what we are referring to when we use the term "cancer." While we often think of cancer as a static state of affairs, it is better to think of it as a process. This leads to a problem, however: when does cancer "begin"? At what point is the disorderly growth of a group of cells a genuine "cancer"?

One might think that the answer is obvious: draw the line at "invasive" growth. In other words, invasion is usually defined as the sine qua non of malignant disease. However, lack of invasion, in some contexts (sarcoma), does not rule out at least some cases as genuine cancers. There is a category of tumors that pathologists refer to as "borderline" tumors ("tumors of low malignant potential" is an equivalent expression). These are examples of a non-invasive cancer that can become very aggressive. Ovarian borderline tumors belong to this category as well. In addition, according to pathologists, there are numerous types of soft-tissue tumors are of "intermediate" malignant potential (Hageman, 2016 Pers. com). In other words, they aren't benign, but they aren't malignant; they're simply intermediate, "benign until they're malignant." They tend to have a very low but non-zero risk of distant metastasis, and the risk exceeds some notional threshold (in the range 0.1–1%). This group of tumors is exceptionally confusing. Solitary fibrous tumors and epithelioid hemangioendotheliomas are some examples. Indeed, some non-invasive cancers turn out to be very aggressive, while some invasive cancers turn out to be only locally invasive. Invasive disease may not compromise organ function or overall survival and reproductive success for some time. Such cases are thus vivid examples of underdetermination: assessment of genuine "function" and "dysfunction" in such cases is ambiguous. But what account of function are the above cases relying upon? Let us consider each account of function in turn.

On an etiological account, all and only traits selected against in the evolutionary past are considered "dysfunctional." To be sure, there was selection in our evolutionary past for the maintenance of differentiation of functional cell types in development and through reproductive age; that is, cells perform different functions in different tissues and organs, and carrying out these functions is relevant (at least over the long term) to survival and reproductive success. So on an etiological view of functions, one might think that any loss of differentiation of cells would count as dysfunctional. But loss of differentiation is a matter of degree, and on the etiological view, dysfunction depends upon how a trait affects survival and reproductive success *on an evolutionary timescale*. Many people survive into old age with pre-cancerous lesions or even slow-growing invasive cancers. So it's not clear that early stages of disease would count as "dysfunctional" on an etiological account of functions. Selection may well have reduced the prevalence of cancer in the very young, prior to reproductive age. So such cases might qualify as dysfunctional on the etiological account. But it is not clear whether cancers after reproductive age would count as compromising function in the etiological sense.

A defender of the etiological view might simply bite the bullet and argue that slow-growing cancers or cancers in old age are not dysfunctional. But this strategy presumes that longevity does not increase inclusive fitness. Inclusive fitness is a measure of how much one contributes to the survival or reproductive success of oneself and one's close relatives. It may turn out that long life may increase the fitness of one's grand-offspring and so, indirectly, one's own fitness. Thus, depending on one's views as to how or whether we ought to consider longevity's effect on inclusive fitness in our application of the etiological sense of "function," cancers in old age might still count as dysfunctional.

There are yet further underdetermination problems facing the assessment of cancer on the etiological account, however. Consider the problem of life-history trade-offs. What enhances reproductive success may compromise longevity. For instance, it may have been fitness-enhancing for males in the evolutionary past to "live fast and die young"—or grow quickly and thus (hypothetically) compete more successfully for mates. However, the very same factors that induce rapid growth and the development of secondary sexual characters—for example, androgenic hormones—may also (as it turns out) lead one to die relatively young from prostate cancer (Crespi and Summers, 2006). Thus, fitness-enhancing traits for one life stage may compromise fitness for another. It is not clear what verdict the etiological view gives for such cases. Indeed, fitness-enhancing traits for one sex might also compromise fitness in the other sex; this is called "sexual conflict." There is some evidence that sexual conflict played a role in the evolution of a variety of traits that both directly and indirectly contribute to cancer risk (Summers and Crespi, 2008). So it may turn out on the etiological view that risk factors for cancer were selected against in one sex and selected for (though indirectly) in another. For instance, longevity may be fitness-enhancing for women, given the provisioning role(s) grandmothers played in our evolutionary past; this is called the "grandmother" hypothesis, meant to explain the longevity of women versus men (Hawkes et al., 1998). On the etiological view, predisposition to diseases of aging such as cancer may well be relatively strongly selected against and so count as "dysfunctional" in women, but less so perhaps in men. Of course, human cultures and sex roles have changed over time. Some anthropologists argue that *variability* and *plasticity* in sex roles, rather than any fixed sex role, were selected for, given the range of environments where distinct life-history strategies might have been optimal (Bribescas et al., 2012). In other words, cancer risk associated with one or another life-history strategy for a given sex may not have been stably selected across all environments. So the question of whether diseases of aging are or are not functional is no easy problem to solve. Perhaps needless to say, it is yet more contested when we have sufficient evidence for a trait counting as adaptive. First, there is the problem of variable environments; second, there is the problem of timescale. That is, whether one adopts a "recent history" account of function or a longer timescale thus may affect whether a trait might count as functional or dysfunctional.

Given these concerns about the contingency and ambiguity of functional ascription on the etiological view, one strategy might be to suggest we consider only traits that enhance inclusive fitness in *current* environments. However, on this view (to be rather crass), living into old age may count as a dysfunction. For at least in the developed world, given the high cost (in terms of both time and energy, as well as financial cost) of caring for the elderly, long life could well compromise the reproductive success and relative survival of one's offspring and grand-offspring, yielding a total inclusive fitness deficit. Whether we take seriously this idiosyncratic (and somewhat tongue-in-cheek) example, the larger point is this: inclusive fitness can vary dramatically across different environments and for different sexes and stages in life history. It seems that whether a state counts as "dysfunctional" in the sense relevant to the assessment of disease should not depend upon these contingencies of selection across variable environments or trade-offs in fitness given various life-history strategies.

Indeed, this is exactly why many who adhere to a naturalist account of disease adopt a consequentialist account of functions. According to this view, it does not matter whether a trait enhances fitness in the short or long term. What matters is a trait's contribution to current survival or reproductive success. Recall, on Boorse's view, that there is a "species-typical" design or norm, departure from which constitutes disease or pathology. Let us turn to objections to this view, and consider how and whether it can address these objections.

2.5. Objections to the Naturalist's View

As it happens, there are four ways in which cancer presents an interesting challenge for the consequentialist, and these dovetail with four major classes of criticisms of Boorse's view. First, there is the "problem of common diseases": given how common some conditions are (e.g., obesity, cavities), how can appeal to the "normal" or average in a population suffice to explain what counts as healthy or diseased? Cancer is, of course, also increasingly common as we age. As discussed earlier, as many as one in two men have slow-growing lesions in their prostate after age 80. Does this challenge the naturalist's view? Not necessarily: Boorse (1977) anticipated and replied to this objection in his original statement. *Actual* statistical averages in a population do *not* determine the norm; rather, the norm is determined by the science of physiology. That is, physiologists investigate and posit normal function and ranges of departure from normal function, granting that actual populations all too often depart from this ideal. However, there is a bit of a puzzle here; if slow-growing prostate cancer is genuinely "abnormal" (at least at the cellular level) but extremely common in old age, ought we to consider it a "disease" or not? The naturalist does not have a clear answer to this difficult question.

The second objection is the "line-drawing problem" (Schwartz, 2007)—this is the problem of how to draw the line between states that are "functional" and "dysfunctional" when there is a continuity between the two, as for average heart rate, blood pressure, or blood sugar level. As mentioned earlier, Boorse's reply to this objection is to simply bite the bullet; he grants that drawing this line will to some extent be a matter of conventional decision:

> Abnormal functioning occurs when some function's efficiency falls more than a certain distance below the population mean . . . this distance can only be conventionally chosen, as in any application of statistical normality to a continuous distribution. The precise line between *health and disease is usually academic*, since most diseases involve functional deficits that are unusual by any reasonable standard. (1977, 559)

It's not entirely clear what Boorse means here by "academic." Indeed, this response seems to leave undecided how we ought to regard such cases. Are early-stage cancers that are subject to variable outcomes genuine disease or not?

The third objection is that it is not clear that biology alone has the resources necessary to provide an unambiguous assessment of species design. Amundson (2000), for instance, points out that functional diversity in humans is a product of widespread developmental plasticity. We have many ways of meeting the same functional goals, and plasticity and variability are indeed general features of humans as a species (for a discussion, see, e.g., Pigliucci, 2001). Amundson thus argues that what he calls Boorse's "functional determinism" is false (the view that there are many uniformly designed individuals and few with novel functional design). He also contends that Boorse fails to acknowledge a difference between the *level* of performance of a function and its *mode* of performance; atypical modes of functional performance are perfectly compatible with high levels of functional ability.

There is a second and perhaps more radical version of this objection. Ereshefsky (2009) argues that it is not even clear that survival and reproduction as a "goal" or ideal is something that can be unproblematically derived from the science of biology. Ereshefsky argues that the naturalist illicitly draws upon a normative assumption regarding goals of human life. To be clear, Ereshefsky does not deny that human beings are the product of evolution; his claim is simply that nothing in the natural world requires that we treat survival and reproductive success as the "apex" or "goal" of organisms. All we may say *as a matter of biology* is that a given trait or physiological state has a higher or lower chance of leading to loss of life or reproductive capacities. Biologists may speak casually in terms of organism's "goals" and "ends," but such talk is simply shorthand for "fitness-enhancing traits"—that is, traits that increase survival and reproductive success. Fitness is a real objective property; calling survival and reproductive success a "goal" of organisms is just a way of talking. I return to this objection later.

The fourth objection to Boorse's view is the "reference class" objection. Kingma (2007, 2010) argues that it's unclear how empirical evidence alone can help us demarcate appropriate reference classes for the assessment of typical or normal function. For a patient with a paracetamol (Tylenol) overdose, liver function is low; low levels of functioning are statistically typical of all individuals in the reference class that includes sufferers of paracetamol overdose. What distinguishes a reference class as "natural"? Kingma argues that such choices of reference class seem to involve an ineliminable value judgment:

> The distinction Boorse draws between health and disease depends on counting only certain reference classes as appropriate. Different reference classes would result in different distinctions . . . Boorse gives no empirical justification for using the reference classes he proposes rather than others; although facts determine both that I am a woman and that I am short-sighted, there are no empirical facts that determine that 'women' is an appropriate reference class, and 'short sighted people' is not. Because the choice of reference classes determines the distinction between health and disease on the BST [biostatistical theory], and Boorse gives no empirical fact that justifies the choice of these reference classes over others, there is no empirical fact that determines the distinction between health and disease on his account. The BST therefore fails to be an empirical or value-free account of health. (2007, 131)

My view of these four objections is as follows. First, I agree with Amundson and Ereshefsky that Boorse *seems* to be relying upon a "natural state" model in biology. This has long been out of favor; species are evolving lineages. The only property that "must" be shared by all organisms in some species (from the perspective of evolutionary theory) is that they are linked via historical ancestry. All organisms vary, and this variation is to be expected and, indeed, is necessary for evolution by natural selection to be possible. Darwin's "population thinking" has replaced the Aristotelian view that what demarcates natural kinds is essential tendencies or "natures" (see, e.g., Sober, 1980). While physiology textbooks might describe the "ideal" heart, as several authors have pointed out (Wachbroit, 1994), such textbook descriptions are idealizations. There is plenty of perfectly healthy, and indeed functional, variation in natural human populations. Amundson is correct to point out that the same high level of functioning might be accomplished by very different modes of organization. When we assign certain physiological states as "typical," we are setting a norm that could be met in different ways, as is vivid to first-year medical students cutting open cadavers.

However, Boorse recognizes and openly acknowledges natural variation; so it is not clear that the fact of variation alone necessarily defeats the naturalist's view. In his "Rebuttal on Functions," Boorse notes: "Now the most obvious logical feature of medical normality is that most functions have a normal range of values. No one value of heart rate, blood pressure, blood urea nitrogen, serum glutamic-oxaloacetic transaminase, forearm strength, height, IQ, and so on is uniquely normal. Rather, there is a range of normal variation around a mean, with either one or two pathological tails" (2002, 101). The mere fact of natural variability and plasticity seems compatible at least in principle with arriving at useful generalizations about the (typical) or statistically average functions of hearts, lungs, and breasts for individuals of a given age or sex. As Hausman has argued, "Evolutionary theory does not imply that physiologists cannot make useful generalizations about the efficiency with which parts of organisms function" (2015, 16).

However, the line-drawing problem and the reference class problem are intertwined in an interesting way. The proportion of individuals we might consider dysfunctional varies with choice of reference class. As mentioned above, cancer becomes more typical as one ages. So what counts as typical or average depends upon the reference class considered. Whether a 75-year-old man with a slow-growing lesion departs from normal function may depend upon a choice of reference class: were we to consider this man's prostate in comparison with that of men ranging in age from 65 to 75, as opposed to 75 to 85, we might arrive at very different assessments. According to Kingma, this is a value-laden choice. Hausman (2011) offers a response: while the biostatistical theory (BST) cannot label as diseases *actual* levels of functions that are normal, the BST can call such conditions diseases because they constitute *diminished dispositional function.*

While this may seem a sound response, Kingma's argument illustrates, in my view, an example of how naturalists and normativists are talking past one another. Naturalism versus normativism about "disease" is not one debate but several. The question Kingma is raising in her argument is not whether one *may* identify diminished dispositional function, but *how and why* we choose to draw the line at one place versus another. Indeed, more recently, Kingma (2014) has characterized four different (or, perhaps, differently nuanced) versions of this debate over value-freedom or value-ladenness in disease designation: "(1) 'health' and 'disease' as *ordinarily used*, (2) *theoretical* or *conceptually clean* versions of 'health' and 'disease', (3) the *operationalisation* of dysfunction and (4) the *justification* for a given operationalisation of dysfunction" (2014, 591). She argues that "these different domains result in at least four possible versions of 'naturalism' and 'normativism', and at least four possible versions of 'naturalist-normativist' opposition" (2014, 591). Kingma's concern is really to do with the latter two.

In the case of early stages of cancer, the issue is exactly this: How and when are we are justified in taking such ambiguous cases as instances of either function or dysfunction? As a matter of fact, this is not simply a philosophical question, but a practical matter that requires attention to the normative goals of medicine, and the best means of enforcing and institutionalizing those goals. Pathologists with different training, at different institutions, using different standards or conventions of assessment arrive at different estimates of function or dysfunction (Welch et al., 2011). These differences may often be traceable to different choices of reference class. In other words, the choice of exemplar representative for a given age or sex is underdetermined by the evidence and such choices are potentially value-laden; a more conservative or risk averse pathologist may choose a different standard than a less risk averse pathologist. Calling this choice merely "academic" hides the normative character of these disparate standards. To the extent that such conventions are moral or normative matters of tolerance of risk, what Kingma is suggesting is that we ought to make these norms transparent.

In other words, once we grant that departure from functionally optimal states is a matter of degree and that assessments of function rely on conventional choices about reference classes, we are faced with a normative question: What conventions ought we adopt in drawing a line between pathology and health? Such choices are arguably based at least in part on precautionary judgments. Reasonable people might disagree on the degree of precaution appropriate in such cases. It seems that even when we pursue as far as possible the empiricist ideal of a naturalistic account of disease, normative considerations may well play an ineliminable role in the assignment of disease status.

The naturalist has a typical response to this line of argument. This argument has confused matters of treatment with problems of diagnosis of disease. All the naturalist can do is identify the disease state, not determine whether it ought to be treated. However, the above discussion made no mention of treatment. The problem is that nature simply cannot tell us (by itself) whether or to what extent function has been disrupted "enough" for something to count as "disease." This is not only because function comes by degree, but also because function is always assessed relative to a specific temporal and spatial scale of analysis, as well as relative to a reference class, and nature cannot tell us which of these is "the" relevant reference class for the assessment of disease.

Hausman's (2012, 2015) account of health as "functional efficiency" to some extent embraces this conclusion. On this view, functional efficiency is a matter of degree. One relativizes assessment of function to the system in which the disruption occurs. "Functional efficiency" of a part or process relative to a given system "depends on the goals of the system." As Hausman grants, however, this characterization of fitness, function, and system identity is *circular*; the function of a "part" of a system is "the effects of the parts that make it more likely that [system] S will achieve [goal] G." Fitness is "how well the part is now able to serve the goals of the systems of which it is a part" (2012, 530). Yet "a system S is one that shows resilience in its pursuit of [goal] G, where that resilience is explained by the special structure of S, rather than by natural law" (2012, 522). In other words, a system is a thing with a structure, and this structure is what contributes to its "resilience" in pursuit of its "goals." A system's goals are defined by whatever effect(s) are promoted by the activity of the parts of the whole. A healthy system is one whose goals are met, which is the same as saying that its parts contribute efficiently to its functioning.

This account makes transparent the system-relativity of functional assessment, and thus why naturalistic accounts of disease as dysfunction cannot resolve *contentious* judgments about how to draw the line in borderline cases; for we must first specify which goal and which system we are interested in and which failure of function. But, it is not always obvious how we ought to demarcate systems, and thus how we assess function. This worry presents itself most vividly in contexts of mental health, because in those contexts it is often rather fuzzy where we ought to locate the "system" that has failed, or the respect in which there is a failure of function. Drawing the boundaries around the "normal" is quite difficult in cases where assessment of function involves assessing "optimal" social or emotional health, adjustment to or achievement of life goals, or other similarly vague outcomes. How much need one's behavior or mood compromise the pursuit of one's goals before we count that behavior or mood as "dysfunctional"? Which "system" exactly is disrupted in mental illness? Given that many mental illnesses involve a complex of causes, affecting a variety of goals and outcomes, how to assess the extent and nature of failure of function is ambiguous.

This problem, it turns out, is not unique to mental illness. Cancer, as we've seen, is a progressive disease that can involve failure in multiple "systems"—from mutations to "cancer genes" that have an effect only at the cell and molecular levels, on up to the extracellular matrix, tissue architecture, blood supply, and response of the immune system, lymph system, metabolic system, and endocrine system, and even (arguably) structural aspects of a patient's environment. There is no value-neutral way to draw the line in assessing what temporal or spatial scale or extent of disruption is either sufficient or necessary for cancer. Rather, we need first to draw a boundary around some system and take just this system to be the relevant one for functional assessment. Such a choice is underdetermined by the evidence; or, nature does not draw the line for us, at least not in any obvious way.

Hausman notes that *any* "directively organized" and "organic" structure persisting over time may be spoken of as more or less "stable" due to the functionally efficient actions of its parts. Thus, he grants that a tumor can be viewed as a more or less functionally "efficient" system with goals. It may be highly efficient, for instance, at evading chemotherapy. As he says: "This account relativizes assessments of functional efficiency to relevant environments as well as to goals of the organism as a whole or to goals of systems within the organism. So it permits one to compare how well the blood vessels within tumors are functioning, even if their functioning is irrelevant to the fitness of the organism as a whole or even detrimental to it" (2012, 534). On this view, there are "healthy" tumors and less healthy tumors, and perhaps also "healthy" subcellular structures and less healthy subcellular structures. Nature does not force us to choose one particular spatial or temporal scale in characterizing systems and their proper functions.

Embracing this consequence of the naturalist account leaves us with an underdetermination problem with respect to many diseases, and particularly those that affect several different systems at once. Consider: at any given time in our life history, chances are that we have one or another "dysfunction" at some temporal or spatial scale, or departure from the norm, in some respect and to some degree. It follows that we are all (to some extent) "diseased," or are all simply more or less healthy. The underdetermination problem gets more complicated still when we consider the problem of overlapping systems and the problem of life-history trade-offs of the sort discussed above—for instance, trade-offs between reproductive fitness and survival. Function and dysfunction may be both disrupted and enhanced in different directions within the same organism at one time or in different time slices of the same organism. The naturalist's account of function and dysfunction gives us very little guidance as to how medical practitioners select (or ought to select) any particular system as of primary relevance to disease assessment.

The naturalist might respond (yet again) that this is merely an "academic" question, a matter of "convention." But this elides, rather than makes transparent, the fact that there is often no way to assess which of many overlapping systems and subsystems and which type of failure of function is most relevant to disease assessment. How much risk to survival or reproductive success suffices to label a state diseased? Where there are trade-offs between length and quality of life, or survival and reproductive success, which ought to take priority? The biostatistical account characterizes such questions as "academic." But, these are exactly those questions that drive genuine controversies over disease assessment.

What the case of early-stage cancer makes clear is that organisms are "kludges" with greater or lesser functional integration, coming by degrees. Some traits promote survival at the cost of reproductive success, and vice versa. The functional efficiency of various traits is often a fragile compromise—a product of trade-offs in life-history optima. Such trade-offs may lead to higher cancer risk; early sexual development in males and height is correlated with prostate cancer risk (Summers and Crespi, 2008). Function and dysfunction are shifting and temporary products of the relationships between systems, subsystems, and their environments and different time slices of the same system. Systems and their boundaries are not fixed features of the natural world, but temporary, fragile compromises. If we grant this, then the functional efficiency of an organism looks less like a single state of affairs for any given sex and age than a shifting and multiply classifiable set of processes and states of affairs. In other words, Hausman's (and Boorse's) account preserves naturalism, but at the cost of hiding the role of values in those cases where conventional choices must be made. When dysfunction comes in degrees, in systems that have at best vague and overlapping boundaries, assessing optimal function is a value-laden decision.

It seems values must play at least some role in the designation of disease, if only in the minimal sense that the "system" which clinicians choose to assess function relative to involves some pragmatic and value-laden choices. By focusing on one subsystem and one goal, clinicians may either fail to preserve or enhance the functions of other subsystems in ways that compromise other subsystems and other goals later in life. Trade-offs in functional efficiency between different parts and life-history stages of any organism means we need to acknowledge that values must play a role in medical decision-making—for we need to decide how to prioritize one function or one time slice of a patient over another.

Consider again the case of asymptomatic cancers in elderly patients. Whether this would count as functionally inefficient depends upon the grain of analysis, or where we pitch our analysis of the system in question, and thus "the goals" we are interested in. In other words, the goals we ought to take seriously are system-relative, and the choice of "system" of greatest relevance to disease assessment is ours. Unless a cancer has progressed beyond the asymptomatic stages, there is no incapacity that the failure to regulate cellular birth and death itself explains, if we understand "capacity" in the sense of organ function. People live for decades with early-stage or indolent cancers. A liver might still be breaking down fats, creating bile, and so on and yet contain a small tumor. In such cases, it is unclear whether we ought to count this as a failure of functional efficiency. On the other hand, an inert tumor in the ovaries of a healthy young woman might compromise fertility. Whether to call this a disease might depend upon which goal we happen to care most about. Which system is the system that should be of interest to medicine?

This is not a given, but a choice, and a value-laden one. While naturalists are certainly correct that there are averages or norms of various reference classes for a variety of physiological states, there is also a great deal of perfectly healthy variation and less than optimal function (cf. Amundson, 2005). On my view, this is not a problem to be eliminated, but a fact that should be addressed frankly and openly.

2.6. The Eliminativist View

The challenges to the naturalist view presented in the preceding section may make the deflationary or "eliminativist" alternative appear attractive. The eliminativist approach eliminates talk of "function" and "dysfunction," and replaces it with state descrptions plus normative judgments. This view was in part inspired by the challenges facing controversial medical cases. Ereshefsky explains:

> Instead of using the terms of 'health' and 'disease' when discussing controversial medical cases, we should explicitly talk about the considerations that are central in medical discussions, namely, state descriptions and normative claims . . . *State descriptions* are descriptions of physiological or psychological states . . . In an effort to avoid normative assumptions as much as possible, state descriptions do not explicitly employ such notions as natural and normal. It may be impossible to eliminate normative elements from many state descriptions in the medical and biological sciences. But at least we can avoid overt uses of such words as 'normal' and 'natural' that often carry implicit normative assumptions. For similar reasons, state descriptions are free of functional claims. The divide between function and dysfunction is controversial, and functional ascription in the medical sciences often carries normative assumptions (Wachbroit 1994, Cooper 2002). To avoid such controversies and assumptions, state descriptions make no claims about whether a physiological or psychological state is functional or dysfunctional.
>
> *Normative claims* are explicit value judgments concerning whether we value or disvalue a physiological or psychological state. (2009, 225)

How does cancer fare on this account? On the one hand, it seems to do eminently well. We avoid many of the difficulties that arose on the dysfunction account; we need not refer to any "natural" or normal state, either in terms of what contributed to our fitness in our evolutionary past or of some "ideal" state of normal function. Moreover, we can acknowledge that state descriptions (e.g., this cell growth exhibits dysplasia) are distinct from judgments about the risk such growth poses or, for that matter, what should or should not be done about such growth.

However, even a relatively neutral state description may carry implications beyond mere description. Ereshefsky notes (citing Okruhlik, 1994, and Amundson, 2005), that many state descriptions carry normative implications:

> State descriptions will never be completely value-neutral, but we can do our best to label value judgments as such when they are identified ... According to the suggestion offered here, we avoid the use of 'normal' and 'natural' in state descriptions. In doing so, we avoid one way that normative concepts get disguised as descriptive ones. We cannot get rid of bias in science, but we should try to eliminate it or highlight it whenever we see it. Naturalist and hybrid definitions of 'health' and 'disease' do not do that. Switching to talk of state descriptions and normative states makes the use of values more explicit. That is an improvement. (Ereshefsky, 2009, 227)

So Ereshefsky's suggestion is that our aim should be to limit value-laden descriptions as far as possible, if only to make more transparent where and how normative judgment genuinely enters in: with respect to whether we regard a given state as desirable or not and whether it ought to be treated. This is, he argues, an improvement upon functional ascription and can help us avoid problems such as prejudging how we ought to intervene.

Ereshefsky is correct to draw our attention to the fact that states don't present themselves as dysfunctional per se; they are judged dysfunctional relative to our interests and our understanding of biology. Early-stage cancer is a vivid example. However, one does not merely describe a physiological state and then make an evaluative judgment. Rather, clinicians make an intermediate judgment about the risk of metastasis or death. Such intermediate judgments require appeal to a body of evidence and theory about the pathological (or genetic, hereditary, or perhaps hormonally determined) character of this particular cancer. Clinicians (and pathologists) who then judge "X" to be a disease state are making a empirically informed bet; they *expect* that an early-stage cancer will cause death, and so they are acting preemptively to prevent the state from progressing. This is to make a judgment of *risk*, one based on certain theoretical assumptions about the progression of disease. Such theories can be false.

In other words, clinicians do not merely describe a state; they make a prediction or hypothesis about what will eventuate. Of course, the objective chance that the disease will eventuate in death may be distinct from the clinician's estimate. The eliminativist thus fails to attend to an important step of clinical judgment, namely assessment of risk based on state description. In the case of early stage cancer, a diagnosis involves not only a state description and an assessment of the desirability of that state, but also an intermediate step, wherein the clinician makes an assessment of risk accompanying such a state (e.g., of mortality, but this is not the only endpoint) founded on some body of evidence. Such evidence empirically informs the normative judgment as to whether that state is sufficiently risky that it should be treated. In fact, there are three senses in which "risk" is involved in decisions about disease status: there is an objective probability (or risk) of a cancer's progression to metastasis, there is the (subjective) estimate of risk, and there is an *epistemic* risk involved in the judgment of that probability of that eventuality.

In principle, whether or not treatment (and what kind of treatment) is warranted does not follow (necessarily) from judgments of classification of a state as diseased. The matter of whether one ought to treat is a distinct normative question, one that should be negotiated between physician and patient. However, in practice, assessments of disease do have import for decisions about care. In other words, *in practice at least*, pathologists' judgments that a given state is diseased are not value-neutral. Such judgments are contentious exactly because they carry normative force, whether in terms of how a given patient is characterized or in terms of practical clinical outcomes. This is, at least in part, why hospitals have tumor boards. Tumor boards are groups of clinicians and pathologists who meet and decide what to do about a patient's cancer. Sometimes, pathology requires interpretation, and this means making a (probabilistic) judgment about what could or is likely to result from a given state. Tumor boards may advise the patient to treat a pre-cancerous lesion aggressively or advise against aggressive treatment. Such judgments involve both empirical and normative considerations.

Ereshefsky's suggestion thus requires supplementation; that is, we should not simply view judgments of disease as descriptive judgments paired with normative ones; rather, we should also view them as paired with assessments of risk. This adds a new, complicating layer to the assessment of disease. Disease judgments are not simply empirical assessments added to value assessments, but theoretically and socially negotiated decisions. The question is not whether they *ought* to be; they are. While, in principle, dysfunction is a matter of empirical assessment, in practice, disease diagnosis involves empirical judgments paired with risk assessments about progression, age, and stage. Any risk assessment requires that we take into consideration tolerance of risk of error; but this is a matter of values.

Clinical assessments involve not only state descriptions and normative judgments, but also linking judgments concerning risk. That such judgments are involved does not mean that assessing whether some state counts as a cancer is entirely arbitrary. There are better and worse assessments of risk, and these depend upon the extent to which one's judgment is informed by the most accurate and comprehensive evidence. Even considering the total evidence, however, the complexity of biology and natural variation means that predicting the future course of some state in a given patient always involves some degree of uncertainty. In the case of early-stage cancer, this means that precautionary judgments play some role.

2.7. Conclusion: From Disease to Risk

In this chapter, we considered two contexts in which uncertainty and epistemic risk come into play: assessing the effectiveness of cancer screening and designating certain early stages of dysplastic growth "cancer." We then turned to the question of how, if at all, competing accounts of disease and health may be relevant to addressing such dilemmas. At first blush, these accounts may seem orthogonal to one another; both naturalists and normativists can grant that functional disruption comes by degree. So at least according to naturalists, this is no threat to a naturalistic account of health and disease. However, what I hoped to show here is that the debates about health and disease and the debates about evidential underdetermination in these line-drawing cases intersect in an interesting way. For what they require is a reconsideration of what the debates over value-laden versus naturalistic accounts of disease are really about. I have argued that there are several overlapping concerns at play: conceptual analysis, challenges in applications or practical assessments of function, whether and how these practical matters are value-laden, and what obligations we have to make such values transparent.

Without specifying whether we are indeed attempting to merely describe, improve upon, or set up normative ideals for medical practice, parties to the debate over disease will continue to talk past one another. Is it wise (or possible) to preserve the empiricist ideal in medicine? What work, really, can a naturalist account of disease (or health) do for us, especially in controversial cases of early-stage disease? Disease assessments are closely tied to assessments of the goals of medicine—at the level of both the individual patient and society at large. But the matter of what medicine ought to achieve is a value-laden question; "effective" care is not simply about reducing mortality, but also about improving quality of life. Arguably, making transparent the ways in which values inform our decisions about disease and health can and should be part of achieving that end. Patients should have input regarding which goals they prioritize. Whether we ought to be exceedingly cautious should depend at least in part upon the values of an individual patient. The overzealous tendency to seek out and identify risk before symptoms are manifest animates the concern of those who have written so passionately about the costs of overdiagnosis and overtreatment (Goetzsche et al., 2012). Given the very serious health costs of aggressive cancer treatment, we ought to be concerned about both diagnosing and treating "cancers" where risk of progression is uncertain. Naturalists' appeal to "convention" or "academic" questions of line-drawing fail to touch this issue. As a society, we need to have a more open discussion about whether investing in screening and early treatment for rare or low-prevalence disease is worth the cost.

In sum, normative judgments seem to play an ineliminable role in disease designation, at least in practice. But normativity per se is cause for alarm only when it is (a) denied or hidden from view and (b) unreflective or (c) founded on pernicious values, such as self-interest (e.g., economic benefit, fear of legal repercussions,). That is, if clinicians are relatively honest about gray areas of benefit with their patients, and if active reflection and consensus are built between patients and clinicians about their values and choices, the fact that judgments of disease status are value-laden should not be cause for concern.

This debate about early stage disease has resonance with larger debates about the aim and practice of medicine. One of the central recent trends in medicine has been a shift to "evidence-based medicine" (EBM) and "precision medicine," which will (on the optimistic view) eliminate bias and found medical judgment on clinical trials of the effectiveness of different treatments or tests. EBM has the goal of directing decision-making in science in a way that is informed by the best available evidence and, in particular, by research founded in clinical trials (for discussion, see, e.g., Howick, 2011; Solomon, 2015). This ideal carries with it both benefits and potential losses. Specifically, it runs the risk of systematically hiding from view the role of norms in medicine, as well as potentially compromising the justified and appropriate role of flexibility of judgment involved in assessing disease status and treatment options on a case-by-case basis. Moreover, by "scientizing" normatively informed decisions, EBM runs the risk of eliminating or denying a role for constructive negotiation between patient and clinician over whether one ought to pursue this or that intervention.

Worrall and Worrall (2001) make a similar argument to that in this chapter in their provocatively titled "Defining Disease: Much Ado about Nothing?" Their thesis is that resolving the long-standing debate about how to define disease will not resolve contentious practical questions about whether to treat various conditions. That is, while such debates may be of philosophical interest, in practical terms we cannot expect such discussions to decide hard cases (Worrall and Worrall, 2001). I agree, in part: it is dishonest to appeal to such definitions as way to resolve normative debates about when and whether we ought to regard a condition as calling for medical care. Addressing such questions requires direct engagement with values. Turning these into debates about whether this or that state constitutes "disease" is a red herring. However, the philosophical debate over this apparently purely conceptual issue does have some practical import. By illuminating how *operationalizing* our concepts of "function and dysfunction" requires careful considerations of choice of reference class and assessment of risk, the debate has helped us see where and how values might play a role in designation of some state as disease.

3

CAUSATION, CAUSAL SELECTION, AND CAUSAL PARITY

WHAT GENES CAN (AND CAN'T) DO

3.1. Introduction: Genes as Causes

Genes (or mutations to genes) are said to "influence," "contribute to," or "increase the risk" of cancer; they "activate," "inactivate," "halt," or "drive" various pathways; they "encode" information; and they interact with various substrates. Causal talk about genes is ubiquitous and varied. In part, this is because genes and their causal roles are varied. One can ask different kinds of questions about the causal roles of genes. One can ask whether genes play *any* causal role in some outcome. Or one can ask how *much* of a causal role they play in some outcome, relative to other factors. Or one can ask *how* specific variants of genes act and interact at the subcellular level (cf. Sober, 2001).

Answering these different questions requires different kinds of evidence. For the most part, one cannot directly observe genes and their action, so one needs to use some proxy measure of the effect of a gene or gene variant. For instance, family histories can provide us with indirect evidence of the role of inheritance, and thus perhaps of genes, in cancer. Gene knockout experiments provide indirect evidence for the specific functional role of a gene. Gene microarrays (where one extracts DNA, RNA, or proteins from a sample of tissue and "hybridizes" them on a chip—or establishes the complementarity of a sequence and probe) are another source of evidence, in this case for the specific causal activity of a specific gene variant. These different ways in which claims about the causal roles of genes are warranted pick out different kinds of causal information, some of which is merely "correlational" and some closer to the outcome of a controlled experiment. What exactly one is gathering evidence *for* using these different methods is not one and the same thing, but ranges from information about probabilistic association to very specific information about the product of this or that linear sequence. Yet all of this information is used in service of claims about genes as "causes of cancer."

This ambiguity of causal talk about genes is complicated further by the fact that what various scientists refer to when they speak of "genes" has shifted quite a bit in the past thirty years, let alone the past century (Burian, 1985; Kitcher, 1982, 1992; Falk, 2000; Fox Keller, 2000; Moss, 2002; Griffiths and Stotz, 2013). Historians and philosophers of biology have argued that the "classical" or Mendelian gene—a discrete physical unit of both heredity and function—is not the same as the "molecular" or the "postgenomic" gene. The combined unit of both heredity and function that geneticists imagined as "the gene" in the nineteenth and early twentieth centuries has been replaced with a far more complex object, or a family of objects, playing distinct causal roles in development and disease and mediated by many extragenomic factors. What we happen to be referring to by the term "gene" in the "postgenomic era" depends upon the context in which the term is used; for scientists talking of "genes" today may have in mind a particular sequence of DNA in a specific region of the genome, several independent sequences on several chromosomes associated with a specific function, or perhaps SNPs (single nucleotide substitutions) (see, e.g., Fox Keller, 2000; Moss, 2002; Beurton et al., 2000; Griffiths and Stoltz, 2013).[1] In other words, the referent of the term is not always the same. While there are precise criteria for identifying "genes" in specific contexts, what one is measuring the effects *of* in these different contexts is not precisely the same thing.

While scientists have precise operational measures for the effects of "genes" *in specific contexts*, when moving from one context to another one can easily make unwarranted inferences. Shorthand ways of talking can lead one to make

---

1. The term "postgenomic" refers to the era after the sequencing of the human genome, which was completed (depending upon whom you ask) in 2001. Since roughly 2001, the understanding of what genes are, how they are expressed, and how extragenomic factors play a role in differential gene expression in different tissues, and at different stages in one's life history, has expanded astronomically. In addition, the scale of analysis has shifted from a focus on "single genes" to genome-wide analyses using "next-generation" or "high-throughput" sequencing technologies. These not only provide fast, cheap, and relatively high-quality information about genome sequences, but enable "annotation" of the genome, or the detection of functional elements of the genome, such as the identification of transcription factor binding sites, promotor regions, etc. In addition, the study of transcription, translation, and epigenetic modification of gene expression has expanded so that it is now possible to explore the distinctions between cancer cells and healthy cells at the genomic level, but also with respect to the role of a variety of epigenetic modifications that play a role in cancer: hyper- and hypomethylation, histone modification, and micro-RNA regulation of gene expression (cf. Lewis and Esquela-Kerscher, 2015).

inappropriately bold claims about genetic causation. Recognizing that we speak of genes and genetic causation in a variety of ways, and on the basis of very different kinds of evidence, can help us see why "gene" talk in biology can lead us astray. In speaking of genes as causes, a biologist may be claiming, for instance, that (1) a particular pattern of inheritance of elevated cancer rates suggests that a mutation to some gene or gene family plays a role in cancer; (2) this mutation plays a specific causal role in a complex pathway to cancer (or at least has been shown to do so in a given experimental context); or (3) variants of a given sequence have been found in five hundred samples of tumors, and perhaps also (4) variants of a particular regulatory sequence play a role in the production of a specific protein in mice, which in turn affects insensitivity to antigrowth signals. Because such claims are made on the basis of very different kinds of evidence, and yet in each case we seem at least to be speaking of the same "thing," claims about the causal role of genes are thus liable to be misunderstood. Overstated claims for the causal power of genes is evidenced by the long and interwoven history of eugenics and medical genetics (Paul, 1995; Kevles, 1998; Comfort, 2014). In part as a consequence of this history of misuse, many philosophers have argued that the concept of "genetic disease" is incoherent at worst or misleading and inconsistent at best (Magnus, 2004; Moss, 2002; Kaplan, 2013; Dekeuwer, 2015). I agree. However, it is also clear that genes do play a causal role in cancer. Indeed, it is fairly typical to hear cancer described as a "genetic" or "genomic" disease. What are we to make of such claims?

Of course, almost all biological traits are affected in some way by genes or genetic activity. In this minimal sense, "everything" is genetic, just as everything is also environmental, developmental, or, for that matter, chemical. Nonetheless, it is not entirely nonsense to speak of genes as playing a distinctive causal role in cancer. The aims of this chapter are to provide a nuanced characterization of that causal role and to use cancer as a case study for a larger family of debates about causal selection and reductive explanations in the philosophical literature. That is, this chapter will discuss what causal role genes do play in the context of cancer, granting that there is a good deal genes "can't do" (Moss, 2002)! I will first provide a framework for talk about causation generally. This will set the stage for an overview of what I call the "mechanistic research program" in the cell and molecular biology of cancer. In my view, this research program has been successful in identifying alterations to pathways associated with cancer, even though mutations to "cancer genes" are only one of many nodes in such pathways and are local, contingent, and highly unstable in their effects.

3.2. Putting Concepts of Causation in Their Proper Place

Philosophical work on causation can be divided (roughly) into three categories:

- Characterizations of the metaphysics of causation
- Inquiry into the epistemology of causal inference
- Debates over the role of causation in scientific explanation

The project of characterizing the metaphysics of causation aims to give a reductive *analysis* of causation in terms of a non-causal concept. Candidate theories are Lewis's (1973) counterfactual analysis of causation, or Hume's "regularity" account. Addressing the merits of these competing accounts of causation is a different matter than characterizing the epistemic conditions on causal knowledge claims. And both such projects are distinct from offering a theory of explanation and the role of causal versus purportedly non-causal explanations in the sciences. Each of these projects has yielded important insights worth preserving, but we need to keep in mind how they are distinct from one another, even if, to be sure, they may be mutually informative.

Hausman (2010) nicely illustrates how such questions have become conflated in the context of debates about one concept of causality, probabilistic causality. He makes some important conceptual distinctions that are particularly relevant to debates about the causal role of genes in cancer; I will draw upon these distinctions in the remainder of the chapter. Hausman argues that theories of probabilistic causation often try to solve several tasks at once—tasks that he argues should be viewed as distinct. They try to offer a metaphysical theory of causation that both sets out adequacy conditions for legitimate causal inference and accommodates probabilistic causal talk in ordinary language and science. But, he argues, causal generalizations such as "smoking causes lung cancer" and claims about indeterministic relations identified in contemporary physics, such as the claim that the collision of a neutron with a uranium-235 nucleus raises the probability that the nucleus will decay, are different sorts of probabilistic claims, requiring different interpretations. So, we should not expect the same concept of causation to be of service in both contexts. That is, what he calls "irregular" causal generalizations—generalizations about causes not invariably accompanied by their effects—are simply different from (and raise a set of epistemic questions distinct from those raised by) claims about probabilistic decay of uranium.

The questions raised by irregular causal generalizations are *epistemic* and *practical* questions, such as when we have good reasons to believe such claims and when it is useful to rely upon them. These matters are independent of *metaphysical* questions about the "fundamental" nature of causation. Attempts to address all these matters simultaneously lead to endorsing claims about the truth conditions of causal generalizations that are, in Hausman's view, unacceptable. I agree; indeed, I will argue that keeping such concerns distinct is key to resolving many of the philosophical debates about genetic causation and "genetic disease." To see why, we need first to make some important distinctions (many of these distinctions or similar ones are also used, e.g., by Woodward, 2003):

- *Causal role versus causal relevance*: A cause can be *relevant* to an effect in different ways: it may increase, or decrease, or have a mixed effect on some outcome. The *way* in which a cause is relevant to an effect is its "role." Information about causal relevance is more general than, and distinct from, information about causal role; for instance, we can know that something is causally relevant to disease risk without knowing whether it increases or decreases risk (or whether it does both, in different background conditions). In many practical contexts, we want to know about causal role, and not just relevance.
- *Variables versus values of variables*: Another way of saying that X is relevant to Y is that there is some functional relationship between the variable X and variable Y, where we can be neutral about the nature of that relation. More precisely, we can say that X can be causally relevant to Y over some range of values, or that X taking one value can cause Y to take a distinct, specific value. Practical causal generalizations (according to Hausman) need not take specific values to be informative or useful. For instance, it may be useful to know that the more cigarettes a day one smokes, the higher one's risk of lung cancer, without yet knowing exactly how much higher.
- *Homogeneous versus heterogeneous circumstances*: Causal factors often have different effects in difference circumstances. When all such background circumstances are fixed, we speak of the causal background as "homogeneous." Most of the time (certainly in the biological world), circumstances are not homogeneous. Generalizations like "smoking causes lung cancer" typically refer to causal relations in heterogeneous circumstances.
- *Types versus tokens*: "Type-level" causal claims refer to causal tendencies; or, such claims are generalizations about how some type of event *tends* to increase or decrease the probability of some type of event. Token causes are actual causal relations between particular events.

Hausman draws upon these distinctions to illustrate how matters of metaphysics, evidence, and explanation have been conflated in debates about probabilistic causation. This discussion is directly pertinent to much of the literature on genetic causation.

To explain, debates about probabilistic causation—and, not coincidentally, genetic causation—often begin by drawing attention to irregular causal generalizations and the "background conditions" problem. That is, smoking (on average) raises one's chances of lung cancer, but background conditions matter. Not all smokers get lung cancer; in fact, most don't. "Type-level" or "causal tendency" causal claims (generalizations about how some type of cause tends to increase or decrease an effect) often "black-box," or ignore, these background conditions. Nonetheless, something can be said to be a prima facie cause of some effect if (a) it is temporally prior to that effect, and (b) the probability of the effect given that cause is larger than the probability of the effect absent the cause (Suppes, 1970, 4).[2] The problem with prima facie causes, however, is that something can be a prima facie cause and yet some confounder or "common cause" could explain the increase in probability of the effect. In order to avoid the common cause problem, some defenders of a probabilistic theory of causation argue that we may say something is a probabilistic cause of something else *only if* it increases the probability of the effect in every causally homogeneous background context (Cartwright, 1979). This has been dubbed the "unanimity" condition (Dupré, 1984). A weaker condition requires only that the probability is raised in at least one context and lowered in none (Skyrms, 1980).

The problem with these conditions on what may count as a probabilistic cause, according to Hausman, is that while they are solutions to one problem, they lead to another kind of problem. Probabilistic theorists of causality misstep, in his view, when they take the unanimity condition, for instance, as providing truth conditions for *any* probabilistic claim of causation *in every context*. This sets the evidential conditions on probabilistic claims of causation so high that it becomes impossible to ever provide useful advice, for according to this view, one could not say that smoking causes lung cancer. One may only say that smoking causes lung cancer only for some populations, in some circumstances. This is because, in principle at least, there are some background circumstances where smoking could decrease the chance of lung cancer (Glennan, 2002) or have little or no effect. To be clear, according to Hausman, the problem with this account is not that this way of speaking does not comport with ordinary language. Rather, the problem is a more serious one: it provides a "theory" of probabilistic causation that attempts to solve too many problems simultaneously. In this way, it conflates (at least) three different questions: what causation is, when causal generalizations are warranted in the ideal case, and when such generalizations are *useful*.

Addressing the metaphysical question of what causation "is," in general, is independent of addressing the epistemic question of when we are justified in inferring type-level "irregular" causal claims about the relative regularity with which X and Y are associated, and these matters are independent of the practical matter of which causal generalizations serve as sound advice. Recall the

---

2. This prima facie causation criterion is similar to Woodward's (2003) "minimal" model of causation.

distinction between causal role and causal relevance. In Hausman's view, practical contexts often *presuppose causal relevance*; their goal is to provide guidance about the *particular causal role*, or value, of the variable in question. Type-level causal tendencies are important to know in practical circumstances. In ordinary life, we know that background conditions are heterogeneous, and so the goal of irregular causal generalizations is to assign a causal role to a variable or identify *causal tendencies*. A claim can be useful and fail to meet an ideal epistemic standard.

The causal role of a variable often depends upon background circumstances, so the truth of causal role generalizations, particularly generalizations about the relative roles of genes—if we are being enormously precise—are always going to be relative to some population or cell or molecular context. But we often wish to extrapolate from one context to another and may not know all the background conditions of potential relevance across the many different contexts. So what we do is see if *on average*, given the widest possible variety of known or potential confounders, x increases the probability of y. Such averaging probabilistic associations are, according to Hausman, the best way to *interpret* causal generalizations like "smoking causes cancer." Indeed, this may be the best way to understand claims like "mutations to *p53* cause cancer" as well. In other words, it's of course not always true that mutations to this gene lead to cancer, nor need it be true that they lead to cancer most of the time for such a claim (properly understood) to be true. All we need to know to assess the practical value of such a claim is that mutations to this gene *on average increase the probability of the effect* (ceteris paribus). This is, in effect, Dupré's (1984) "average effect" account of probabilistic causation (see also Hitchcock, 1998).

These distinctions between causal role and relevance, values of variables, type and token, and so on are particularly important to keep in mind when we turn to causal talk in biology and medicine and, in particular, in the context of talk of the causal role of genes. First, it is important not to confuse claims about causal relevance with claims about specific causal *role*. To say that genetic mutations are relevant to cancer is not to make a commitment (so far) to the matter of what specific causal role they play. Second, we speak of causal relevance (e.g., raising relative risk) without yet addressing questions about both the "stability" and "invariance" of such generalizations or, for that matter, specificity of causal role (Woodward, 2010). Knowledge of how common it is for causes of a certain type to be associated with a given effect is useful for various purposes, including both practical and scientific ones.

Practical causal generalizations that make claims about causal relevance, such as "smoking causes lung cancer," play important roles in the biological sciences and public health. Such claims are best interpreted (Hausman claims) as assertions that a type of cause increases an outcome's probability *on average*, relative to a population absent such a cause (where such populations are assumed to be otherwise relatively uniform with respect to asbestos exposure, nuclear fallout, etc.). Appeals to such "irregular" causal generalizations may fall far short of ideal epistemic conditions on probabilistic claims. Nonetheless, they are one of the few options open to us if we wish to generalize about causal relations in a world where causal backgrounds are heterogeneous. Indeed, the enterprises of epidemiology, public health, and clinical medicine presuppose that our speaking of such general tendencies is warranted. For instance, the surgeon general may be interested in citing such facts in giving warnings about the effects of smoking, or clinicians may wish to recommend taking aspirin after a stroke. We need to keep separate the practical matter of whether causal tendency claims are (generally) useful and the question of whether they meet ideal epistemic conditions.

The claim that mutations to *p53* "cause cancer" may thus be interpreted as a shorthand for saying that the protein produced by this gene (with, of course, help from the machinery associated with gene transcription and translation) is associated with arrest of cell division, so that mutations to this gene can (provided many other conditions are in place) raise the risk of cancer. Were every claim in a textbook in cell and molecular biology of cancer as precise and carefully circumscribed, the text would be far more accurate, but very difficult to read!

In science, it is often necessary to give causal "sketches" or shorthand characterizations of enormously complex and context-sensitive causal pathways, either for pedagogical purposes or for ease of communication. Unfortunately, philosophers entering into such conversations sometimes call upon ideals of evidential reasoning that may be impossible to meet in practice or inappropriate for the purposes of inquiry. This is why detailed attention to the practice of science is so important; our ideals must, of course, be practicable, as well as appropriate to the context. While in some cases it may be possible to know that, given a specific cause, a specific effect may come about without exception, more often, in biology at least, at best we can know how X intervention yields some outcome Y for the most part, or in model systems that approximate our target system only to some extent or in some respects (Steel, 2007). In such cases, it is perfectly sensible to talk of causal role and causal relevance without necessarily specifying all possible boundary or background conditions (cf. Suppes, 1970, 7–8), provided that we are clear about what we mean. There, of course, is the rub.

In this chapter, I bracket questions about the metaphysics of causation and concern myself primarily with epistemic conditions on claims about the causal relevance, causal roles, causal tendencies, and causal specificity of genes in the context of cancer. My argument is as follows. While philosophers have been right to critique overly reductionist or "gene centric" characterizations of cancer causation, we need to be very careful to keep separate questions about actual causes and questions about causal tendencies. We should beware of applying epistemic conditions of adequacy for the former to the latter; for scientists are interested not simply in actual causes or relationships between particular events, but in causal generalizations and their relative stability. The conditions of adequacy for evidential claims about such distinct causal claims are different. This is connected to a larger argument in this chapter: the "causal selection" problem is not one problem but several. When scientists "select" causes as most "important" or worth attention, they may be seeking different kinds of information: (a) whether X types are causally *relevant* to some general outcome (or type of outcome), (b) whether X is an actual cause of some specific outcome, (c) what specific causal role X events play in some type of outcome (and relatedly, perhaps, how to effectively intervene), or (d) how stably related X events are to outcome events of some type. Which kinds of causal information we wish to have will determine which evidential conditions are appropriate. Philosophers have (in my view) mistakenly taken the "causal selection" problem to be one problem. But how and when we are warranted in taking a cause as significant or explanatorily important is not a single problem but depends upon what we wish to do with this information, such as which kinds of interventions or explanatory generalizations we wish to make. The mistake philosophers have made, in my view, is hoping for a general, systematic answer to the "causal selection" question; this mistake will be less easy to make once we appreciate the variety of causal information sought, or causal selection problems, in the sciences. The focus on genes as causes of cancer was a deliberate methodological strategy taken by twentieth Century cancer researchers (cf. Fujimura, 1996). It was a successful strategy, though one that surely relied on reductionist heuristics. Let us turn to what I call the "mechanistic research program" in cancer research.

3.3. The Mechanistic Research Program in Cancer

The search for mechanisms for cancer at the cell and molecular levels, or "cell-intrinsic mechanisms for" cancer dominated the latter half of the twentieth Century. I will call the hunt for mutations that lead to disruptions of the cell's typical functions via alterations of mechanisms that typically regulate cell birth and death the "mechanistic research program" in cancer research. This research program's focus is on disruptions of these internal regulatory mechanisms, disruptions that lead cells to continue to divide and grow. This picture of cancer is often associated with what has come to be called the "oncogene paradigm" (see, e.g. Fujimura, 1997; Morange, 1993, 1997; Moss, 2002; Bertolaso, 2016).

In what sense is the oncogene paradigm a "mechanistic" research program? It may be helpful to start with the concept of a mechanism. There are a variety of competing characterizations of mechanisms, but very broadly speaking, a mechanism is an arrangement of parts, operations, and activities that produces a set of effects in a regular way, typically in the production of a function. Typical biological mechanisms are those that control the replication of DNA, photosynthesis, or the firing of neurons. Each mechanism is organized; that is, there are distinctive "start points" and "endpoints" and a productive organization of entities and activities that regularly produce some outcome. Mechanistic science is reductionist, in the very thin sense that such research involves decomposing a system and characterizing the activities of its parts and processes. Characterizing a mechanism serves a variety of roles: it identifies the parts and processes that yield a specific causal outcome and thus "explains" the outcome, or identifies the causal relations relevant to the production of the phenomenon under investigation. According to Bechtel and Arahamson (2005), explanatory understanding often stems from the process of simulating how a mechanism's components are organized and interact so as to produce the phenomenon to be explained.

Mechanistic inquiry has a very attractive property: the discovery or characterization of the mechanism for "X" describes "how it works"—that is, these parts and activities lead to these regular outcomes in these circumstances. Talk of "how it works" carries several implicit assumptions: (a) that this is the "main," "essential," or "fundamental" mechanism "for" cancer and (b) that mechanistic understanding is "portable"—that is, that the system and its parts and activities will behave in the same way in different circumstances. Identifying such portable mechanisms satisfies a particular ideal or model of scientific explanation, one tied closely to the hope for (local) intervention. In fact, the oncogene research program was promoted by its advocates as an effective tool for intervening on cancer (Fujimura, 1997), even though its advocates have become increasingly aware that the chain of causation was enormously complex. The mechanistic ideal in cancer research takes the identification of specific causal roles of genes or proteins as central to explaining cancer. Mechanistic explanations are, at least in principle, an improvement over "mere" correlations; for with a mechanism in hand, one can show not only *that*, but also *how*, cause and effect are related. Mechanistic explanations are thus taken to meet a variety of ideals jointly: they facilitate understanding by decomposing systems into parts and identifying the specific causal roles of these parts, insofar as these mechanisms are "regular" or "portable"; they play a (potential) interventionist role, in identifying specific ways in which (at least in principle) intervention on a specific outcome might be achieved; and last but not least, they play a justificatory role in pharmaceutical development and regulatory policies.

As we saw in Chapter 1, however, drawing boundaries around mechanisms and characterizing inputs and outputs often involves pragmatic choices: picking out specific outcomes of interest and tracing specific causal pathways to that outcome. The more we learn about mechanisms of regulation of the cell cycle, the more arbitrary such choices may seem to be. For it turns out that the "boundaries" around a mechanism, even for a relatively precisely specified outcome, are often vague: we need to decide how many parts of a network of causal interactions are of relevance, and of relevance to which of many possible outcomes. When, how, and in what order core parts of a mechanism are activated may be contingent on a variety of "upstream" factors, some more remote in time or farther out in space than others. This is not to say that mechanisms regulating the cell cycle are not "real" or do not play important explanatory roles in cancer. Rather, the claim is that characterizing a mechanism involves a choice and the choice is not "obvious." Cells divide or fail to divide in response to a massive array of both intra- and intercellular signaling pathways. Which pathways are part of the "mechanism for" apoptosis, for example?

The mechanistic research program in cancer research, at least for the past twenty-five or thirty years, has aimed to solve this problem by identifying and characterizing only one pragmatically bounded subset of a wider system of regulatory controls and causal pathways that play a role in cancer, cancer cells. But, of course, when and whether a particular "mechanism" plays a role in any particular outcome depends upon the many contingent additional facts: not only the timing and nature of the compromise to a particular bounded pathway, but also what other pathways have likewise been compromised, available blood supply, immune response of the patient, and much else besides.

By way of example, Weinberg (2014), in his textbook on the molecular biology of cancer, speaks of the cell cycle clock as the "clearinghouse" and "master governor" of the cell nucleus, a "machine" whose proper function is zealously guarded by "surveillance mechanisms": the checkpoints that play a role in advancing or halting the cell cycle. Cancer cells, he says, suffer from a breakdown in the "circuitry" that controls the cell cycle:

> The cell cycle clock is a network of interacting proteins—a signal processing circuit—that receives signals from various sources both outside and inside the cell, integrates them, and then decides the cell's fate. (276)
>
> Like virtually all machinery, the machine that executes the various steps of the cell cycle is subject to malfunction. This fallibility contrasts with the stringent requirement of the cell to have various phases of the cell cycle proceed flawlessly. For this reason, the cell deploys a series of surveillance mechanisms that monitor each step in cell cycle progression and permit the cell to proceed to the next step in the cycle only if a prerequisite step has been completed successfully. (279)

This way of representing the "machinery" responsible for cancer involves focusing on a particular grain or temporal and spatial scale of analysis, and usually a very narrow or specific set of features of cancer cells. On this picture, cancer cells are "autonomous agents," acting alone. Weinberg's textbook identifies a discrete function of the cell (e.g., apoptosis) and then decomposes the function into specific activities (cell cycle arrest) involving specific entities (genes and proteins) and processes (upregulation, downregulation of production of protein, blocking of a protein binding to a downstream regulator of a further protein, etc.), all of which contribute to apoptosis. One such entity is *TP53*. Weinberg characterizes *TP53* as the "guardian and master executioner" of the cell; it plays a central role in halting cell division, or preventing progression through cell division, and initiating apoptosis (cell death). Likewise, *pRb* blocks advance through the cycle. There are several such so-called driver genes in cancer that are associated with core regulatory pathways in the cell, perturbation to which affects whether cells can continue to grow absent signals from the tissue microenvironment or be resistant to apoptotic signals.[3]

TP53 is a gene that ordinarily produces a "cellular alarm protein," causing a cell to enter into a quiescent state or apoptosis (cell death), via its ability to induce expression of *p21*, which in turn produces proteins that arrest the cell cycle. Thus, a mutation to either *TP53* or *p21* or to both can lead to cellular proliferation because of failure to halt the cell cycle. This is an instance of a mechanistic explanation of one of cancer's core hallmarks—uncontrolled proliferation. However, we have given at best a cursory description of this pathway, and it is only one of many possible ways in which uncontrolled proliferation comes about. Mutations to *TP53* are only one in a set of overlapping signaling pathways that can lead to (or inhibit) proliferation. Thus, while we can identify the role played by this mutation in one context, with a specific outcome, how we demarcate the boundaries of the mechanism is a pragmatic matter.

Much of the mechanistic research program in cancer research has involved identifying so-called cancer genes and their causal roles in mechanisms controlling the cell cycle. Some mutations are said to "promote" cancer; some might be better said to "allow" cancer. "Oncogenes," when mutated, are said to lead to the overproduction of growth-stimulatory signals, which *can* lead to proliferation typical of cancer cells. "Tumor suppressor" genes play the opposite role: they constrain or suppress cell proliferation. The antigrowth signals they produce prevent cells from growing without limit. For instance, *APC* is a "tumor suppressor" gene. It plays an important causal role in the degradation of beta-catenin, which is an important regulator of cell division. Beta-catenin is expressed at high levels in the interior of the colon crypt, pockets of cells lining the colon that contain "stemlike" or continuously proliferating cells at the base and more differentiated

---

3. Though some (Tomasetti et al., 2015) complain that the term "driver gene" has taken on a shifting meaning, i.e., any gene that plays some major causal role in the progression of cancer (though how great a role is uncertain).

cells as one moves up to the surface. At high levels, beta-catenin helps maintain the "stemness" of colon stem cells. It enables them to continue proliferating. These cells proliferate, differentiate, and migrate up to the surface of the colon to form the intestinal lumen—the epithelial lining of the gut. Ordinarily, as they migrate up to the surface of the colon, the cells lose their "stemness" because *APC* triggers the production of a cascade of proteins that eventuate in the degradation of beta-catenin. When the *APC* gene is inactivated, beta-catenin accumulates at the surface of the crypt, yielding polyps in the colon. Thus, individuals who inherit these mutations or who acquire them have excessive development of polyps. The gene is called a "tumor suppressor" because of its active role in suppressing the overgrowth of cells.

Genes like *APC*, *SRC*, and *TP53* are often represented in textbooks of cell and molecular biology as nodes in gene regulatory systems, where the images bear a deliberate resemblance to a "circuitry" with controls that halt or accelerate activity in the cell. This resemblance is no coincidence. The picture of the cell as a circuit is one inherited from the advances in cybernetics and information theory in the mid-twentieth century.

But how does the regulatory pathway control cell division like (and unlike) a circuit? To begin with, what are circuits like? Circuits come in (roughly) three flavors: series circuits, parallel circuits, and combinations of the two. Series circuits are the simplest case: a conductive path connects a power source to the circuit and passes through a resistor, an object that uses the electricity to do work—this could be a lightbulb or an electric motor. In parallel circuits, there are multiple pathways along which a charge can travel to separate resistors; each resistor is placed in its own separate *branch*, and the branching points are called "nodes." A charge passing through the loop of the external circuit passes through a single resistor present in a single branch and can "decide" to return to the power source or not. Combination circuits include both series and parallel pathways. Circuits are typically represented by a simple set of symbols. A conductor (connecting wire) between any two components of the circuit is represented by a straight line. A resistor is represented by a zigzag line, and an energy source is represented by a series of parallel lines. A "switch" can interrupt the flow of charge; an open switch is generally represented by a break in the line (see Figure 3.1).[4]

Diagrams of cell regulatory pathways often bear a deliberate resemblance to circuit diagrams in a number of respects. They are represented as akin to a series circuit—with a circular pathway, along which signals or information are

---

4. There are other, more complex variations on this theme. For instance, amplifiers modulate the energy passing through a circuit by drawing upon a power supply, e.g., to increase the output. An amplification circuit receives a signal from the environment that amplifies the energy passing through the circuit. (Thanks to my father for explaining circuit design.)

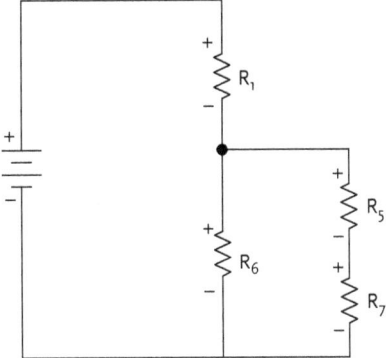

FIGURE 3.1 A simple circuit diagram.
From *Wikipedia*, "Electrical Circuit," accessed June 1, 2015. http://simple.wikipedia.org/wiki/
Circuit_diagram

said to pass. There are nodes in these pathways that play functional roles akin to those in a circuit. Some nodes "interrupt" the flow of information; these are analogous to switches that are "open" in that they "block" or "inhibit" activity or the continued production of a signal. Other nodes serve as "resistors"; they "do work" with the information—usually by directing the signal to activate another pathway, which may in turn either induce transcription of a gene or "upregulate" a protein. Indeed, at the close of his chapter on apoptosis and *TP53*, the "Master Guardian and Executioner," Weinberg includes a representation of the "apoptotic circuit board" (see Figure 3.2). The diagram is rather atypical in that it deploys symbols deliberately intended to mimic those in a circuit diagram. The symbols have different meanings from those typical of a circuit, and there are many more types of activities and states at work in Weinberg's cell circuit diagram than are typical for a circuit. For instance, rather than a single line indicating a current, activation is in blue, inhibition is in red, transcriptional regulation is in green, and so on.

Granting the pedagogical purposes of such diagrams,[5] even the most simple cell signaling pathways are distinct from circuits. They are often "activated" by what I call "floating intervenors": external signals with more or less specificity (e.g., "IGF-1," "cyclins," "ligands," "mitogens," "DNA damage," or "cellular stress"), the sources of which may or may not be indicated in the diagram itself. This is because there are no "closed" circuits in the cell. Or perhaps it is better to say that closed circuits are only those whose connections to other circuitry have been temporarily blocked. Moreover, most cell pathways have multiple

---

5. And the shared history of cybernetics and the growth of genetics (see, e.g., Sarkar, 1996; Griffiths, 2001).

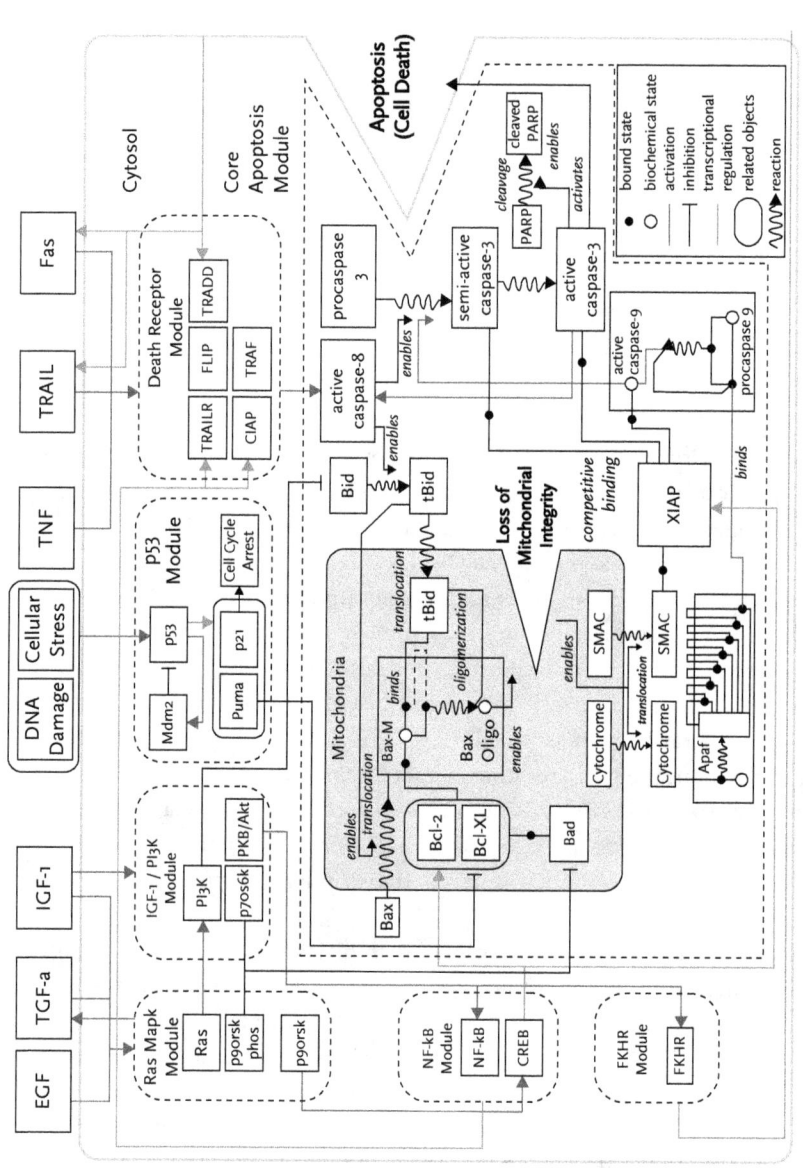

FIGURE 3.2 The apoptotic circuit board.

From R. Weinberg, *The Biology of Cancer*, 2nd ed. (New York: Garland Science, 2014), fig. 9.40.

"inputs" and more than one output or "endpoint." Though, for ease of understanding, regulatory pathways are represented in textbooks as if they have single endpoints. But the same elements in a pathway ending with X may play a variety of functional roles in regulatory pathways that result in Y, Z, and (it may well turn out) yet a third byproduct that in fact *inhibits X*. This is, in fact, exactly why these pathways are often represented as "circuits," insofar as they are often self-regulating or "multiple operations may be performed simultaneously and individual components may have effects on other components responsible for other operations"(Bechtel, 2015, 201). That is, the very same pathway may be deployed for different purposes, drawing in different elements at different stages in the development or the life cycle of a cell and depending on the context. Finally, many pathways coregulate one another. In other words, the causal role of genes is context-dependent. Cancer cells are not autonomous agents.

What cell and molecular biologists refer to as "mechanisms for cancer" are the entities, processes, and causal interactions that yield specific types of failure of function at a local scale—failures associated with regulation of the cell cycle or cell birth and death, as well other capacities or hallmarks of cancer cells: the ability to attract a blood supply, evade immune destruction, and reprogram metabolism of the cell. But as we've seen, mechanisms for specific capacities of cancer cells are "activated" in context-sensitive fashion: they are local in their effects and highly unstable. For example, the same mutation (or epigenetic change) can have different effects, depending not only upon which cell type and tissue of origin it occurs in, but also upon tissue architecture, sex or age of the patient, and even timing of onset. The same mutation can have different effects on the progression to disease, depending upon whether it arrives early (e.g., in the first dysplastic cells) or late (e.g., in the context of an advanced heterogeneous population of cells in a tumor).

Features of cancer cells are only one part of the complex process yielding cancer. That is, cancer, like human behavior, is a process of production of locally activated dispositions of cells and tissues, dispositions that are highly sensitive to context and subject to interactive effects. Understanding patterns of organization in these interactions is one part of understanding cancer progression. The mechanistic research program in cell and molecular biology has been enormously successful, but it is only one part of a complex story. We can also "zoom out" and place these cell and molecular mechanisms in larger environmental, developmental, and evolutionary contexts or take a multiscale or "systems" approach. These are all ways of placing these context-dependent, local, and highly unstable causal mechanisms in a larger context, and thus better appreciating their causal roles in cancer.

For instance, multiscale models integrate "bottom-up" experimental data documenting the behavior of particular genes and proteins and their dynamics with "top-down" approaches. The mechanistic research program in cancer research in the past two decades has undergone a gradual shift from focus on the local activities, entities, and processes to a systems-level approach that attempts to integrate this local data and find patterns across different "networks" (see, e.g., Green, et. al., 2018).

Putting these mechanisms in evolutionary and developmental context helps to make sense of why so many redundant regulatory pathways control cell fate. To explain this point, I will draw upon Wimsatt's ideas concerning "generative entrenchment," Burian's notion of "molecular epigenesis," and Woodward and Dupré's discussion of "robustness." First, Wimsatt and Schank (2004) describe "generative entrenchment" as a "dynamic structural feature" of certain kinds of "generative structures." The structures they have in mind are those involved in phenotypic development, such as gene regulatory networks and even "gene, morphology, and behavior pairings with environmental stimuli." That is, any system that has organization and regular behavior and that develops over time could have more or less generatively entrenched features. Thus, insect colonies or even institutions, such as scientific regulatory bodies or religious communities, are structures whose parts could exhibit more or less generative entrenchment. A system has generative structure when it

> has a characteristic set of causal interactions which could be variously represented. One of the simplest representations is a directed graph, where nodes are parts, processes, or events, and arrows are consequences of the presence or operation of nodes on other nodes. For each node, consider how many other nodes can be reached from it by following the arrows. This indicates how much of the phenotype is downstream of, causally dependent upon, or affected by a given node. We define the generative entrenchment (GE) of a node as the magnitude of its downstream dependency. (Schank and Wimsatt, 2004, 360)

That is, generative entrenchment is a feature of generative structures, or structures with complex causal pathways (like gene regulatory networks). A node of such a system is "generatively entrenched" when it has many downstream effects. In biology, such a node may be likely to be of great adaptive or functional significance, because changes to these nodes have multiple downstream impacts—for instance, not only when changes in the value of a specific variable play a role in producing a specific protein, but also when that protein plays a key role in essential stages of development, fertilization, wound healing, or placental implantation. *p53* leaps to mind as a vivid example.

Schank and Wimsatt, drawing upon Simon (1962), argue that evolutionary processes tend to give rise to generative structures. Core developmental processes with multiple downstream effects are likely to be highly conserved across a variety of organisms arising from a common ancestry and to arise earlier in development (Carroll et al., 2013). Highly entrenched features tend to be evolutionarily conserved, because a loss or compromise to such features will have many downstream effects or cause serious compromises in fitness. There is a tension in evolution, however, between conservation and evolvability. As Simon (1962) has argued, relatively modular or independent "stable subassemblies" have the following advantage: a system whose parts develop and vary independently without affecting the function of other parts is more "evolvable," where this is understood as being better able to generate adaptive novelties that may in turn be selected. Similarly, Lewontin (1979) has argued that more evolvable lineages are likely to have distinctive "quasi-independence" of parts. Schank and Wimsatt say this leads to the "GE paradox":

> How can complex adaptive systems evolve and continue to evolve in any other than a predominantly accretional way if their generative elements become increasingly entrenched with increasing complexity? How does this permit continued modular evolvability? (2004, 363)

That is, paradoxically, there are fitness advantages due to entrenchment, and fitness advantages due to modularity, but the two cannot be maximized simultaneously: "Modularity, duplication, and functional redundancy should each *decrease* entrenchment by reducing interdependence between or dependency upon specific system components" (Schank and Wimsatt, 2004, 363.) In sum, generative entrenchment and modularity are features of systems that can vary continuously, and a structure with highly entrenched parts is (in general) less modular. But both features have fitness advantages over the long term. Thus, the question is how, over the course of evolution, these competing features of developing systems could be optimized simultaneously.

The answer is: they cannot. In fact, cancer illustrates exactly how entrenchment and modularity each have fitness advantages and fitness costs. Moreover, the fact that we are evolved and thus subject to these trade-offs explains in part why a purely mechanistic research program in cancer research needs supplementing with a multiscale approach. Recall from the earlier discussion of *TP53* that it is part of a mechanism for cancer progression. But a single mutation to *TP53* does not (necessarily) result in a cancerous cell. While mutations or failures in entrenched nodes (like *TP53*) are certainly bad news, one mutation does not suffice for cancer. This is in part because of the modularity of tissue structure and architecture, and because of the variety of "redundant" or "backup" regulative pathways that can prevent failures in even highly entrenched nodes like *p53* from leading to disease. So, in order to understand why mutation to one or two genes alone is insufficient for a cell to become a cancer cell, we need to understand the trade-offs between entrenchment and modularity in complex evolved systems, and the role of redundancy in protecting us against disease.

Equally entrenched preventive barriers to cancer are epigenetic controls on gene expression. The term "epigenetic" can be used in both a narrow and a broader sense (cf. Griffiths and Stotz, 2013). "Epigenetic" in the narrow sense used by most cancer researchers refers to mechanisms that regulate gene expression via alteration in chromatin structure (without altering the DNA sequence itself). Chromatin binding affects which genes are active and which are repressed in the cell. DNA is usually wound tightly into a chromatin complex in ways that affect patterns of gene expression. Active genes have "open" chromatin—they can be transcribed, due to the fact that the DNA is "unwound" from the histone "spool." This is called DNA hypomethylation. In contrast, a "repressed" gene has closed chromatin, or is hypermethylated. Narrow-sense epigenetic modifications to DNA include genomic imprinting, X-chromosome inactivation, or silencing of repetitive elements. Some of these modifications occur early in development, as a normal part of development. Cancer progression is affected both by genetic and by epigenetic modifications to the genome; that is, without a permissive epigenetic environment, so-called cancer mutations will not lead to cancer. In a suppressed gene, a mutation will not cause harm. That is, cancer is as much an "epigenetic" as a "genomic" disease. Alterations to methylation can also yield cancer; entrenched features that enhance fitness can also have a fitness cost.

Such narrow-sense epigenetic alterations are not the only "permissive" requirement for cancer. The tissue microenvironment also plays a key role in whether and how often populations of cells advance to invasive metastatic disease. Gene expression is highly contingent on the cell and molecular environment; these contexts of gene expression determine in large part which genes are transcribed, when, and how. Burian calls this set of regulatory controls on gene expression "molecular epigenesis":

> A cellular context is required for DNA to function, and different cellular contexts extract different information from the same DNA sequence. Furthermore, the pathways and networks into which nucleic acids and their products enter are multi-leveled and are replete with feedback loops that cross multiple levels. Yet further, the physiological and nutritional states of cells, exogenous signals from the extracellular matrix, other cells, or the external environment . . . alter the networks and can have stable molecularly epigenetic effects with dramatic lifelong consequences for the organism's morphology or physiology . . . Signal transduction modules work as packages, but what they do—what gets transduced to what by a given signal transduction module—is affected by evanescent signals,

physiological states, chromatin packaging, timing, temperature, integration of the signal into a variety of larger modules, and much else. Thus the networks have components of strikingly different sorts, and their behavior is affected by intra-cellular, extra-cellular, and external environmental conditions. (2004, 63)

Burian's observation is true for functional cells, but it is no less true for cancer cells. Even a cell with all the typical mutations associated with cancer cells will not behave as a cancer cell without cooperation of the tissue microenvironment. Advanced tumors are a complex of cancer cells and supporting stromal (normal) cells, such as fibroblasts. The stroma plays an essential role in the progression of a solid tumor, by providing a structure in which the tumor can grow and by mediating processes essential to cancer progression: angiogenesis and, eventually, metastasis. In other words, the tissue microenvironment may be more or less permissive in the progression to metastasis. This is why understanding the cell-intrinsic mechanisms that affect the behavior of cancer cells is only part of explaining the progression of cancer.

In order to put these cell-intrinsic mechanisms in their larger context, we may also appeal to the idea of "robustness." Cells and cell populations may be spoken of as more or less "robust" in the following sense: they have properties that enable them to persist in their functionality, despite perturbation of either environment and/or internal structure, organization, and/or genetics. The properties of "robustness" have been attributed to gene regulatory networks and to populations of cells (Kitano, 2007a). A network may be more or less "fragile" and more or less "robust," in the sense that it may be highly resistant to various environmental insults. Thus, knocking out a single gene or acquiring a single mutation may or may not affect any given network if it has a high degree of redundancy built into it or ways to "find solutions" to pathways that overcome the losses or failures due to a mutation in a single gene. Some argue that more heterogeneous populations of cancer cells are more evolvable; others argue that it's the hierarchical structure of cancer cell populations—the division of labor of cancer stem cells and non-stem cells—that makes them more or less likely to become an invasive tumor. Others argue that it is the interactions between cancer cells and their tissue microenvironment that is essential to the relative robustness of a tumor. In fact, it's possible (and likely) that all such factors play a role. That is, the behavior of a cancer cannot be explained solely in terms of the aggregative properties or mechanisms at the cell and molecular levels, but requires attention to the dynamic interactions with tissue architecture, to the relative modularity and robustness of different mechanisms that protect disorderly growth, such as tissue architecture.

In other words, cancer cannot be explained from the "bottom up," strictly in terms of the genetic and molecular features of cancer cells, alone. This is the intuition behind some "systems" biological approaches to explaining cancer; their goal is to seek systemic structural properties of signaling networks that make cancer more or less likely. If there are stable "robust" features of such networks, discovering such features may be a tool for enhancing our understanding of how and why we are vulnerable to the disease and how we might intervene to prevent or halt the course of cancer progression. This research program is still in its infancy, but there are a variety of models of the evolution of robustness in cancer (see, e.g., Kitano, 2004, 2007b; Huang et al., 2009).

What is distinctive about this research program? There are a few distinctive features of systems approaches to cancer (see also Brigandt et al., 2016). Dupré and Woodward (2013) speak of "emergent" or "robust" systems as well as systems or network-level explanations in biology. We can speak of both healthy tissues and cancerous tissues as more or less robust systems, in the sense that they persist or succeed in regenerating themselves despite the environmental challenges presented by the immune system and larger paracrine signaling systems that surround them. The reason most proto-cancers fail to progress is that such systems evolved to repress the "autonomy" of cancer cells in the body. The outcome of any particular genetic mutation is highly contingent; genes and the mechanisms of which they are a part are local, context-specific, and highly unstable.

Reductive mechanistic decompositions of cell-intrinsic capacities associated with cancer at best provide local explanations of specific outcomes. Mechanistic, local, reductive accounts have played a significant role in explaining cancer progression. But a piecemeal articulation of each hallmark or capacity of cancer and its associated mechanisms targets only one temporal and spatial scale and one family of outcomes in cancer progression. In order to piece together how cancer emerges as a process, we need to attend to the interactions of these cell-intrinsic and cell-extrinsic signaling pathways, operating at a variety of scales, and their dynamics over time. Over the course of development of a cancer, such networks evolve and coevolve; understanding cancer requires not simply understanding each failure of mechanism in isolation, but the dynamic processes of feedback between them. With this overview of the mechanistic research program in cancer in hand, as well as a discussion of its scope and limitations, we are in a better position to disambiguate talk of cancer as a "genetic disease" and the "causal selection" problem.

3.4. Causal Selection, Causal Parity, and Genetic Disease

Consider Figure 3.3. On the y axis are the major causes of cancer. On the x axis are the estimated number of deaths in the United States attributable to that cause. The image describes "realistic goals for reducing cancer mortality"; that is, it describes ways in which we might intervene on various causes of cancer so as to

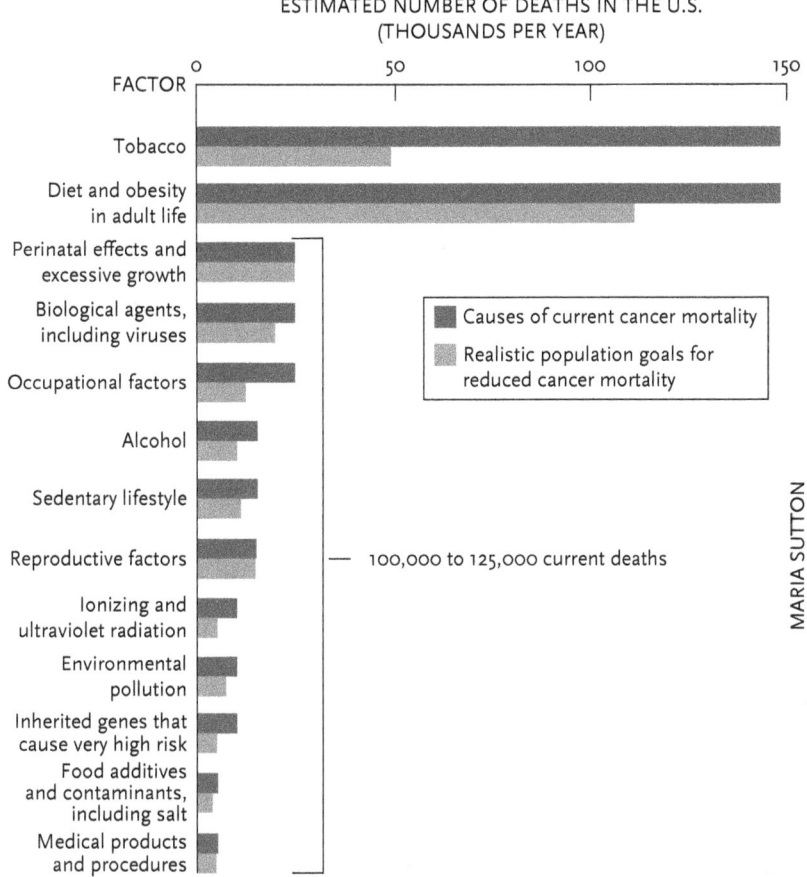

FIGURE 3.3 Strategies for minimizing cancer risk.
From Walter C. Willett, Graham A. Colditz, and Nancy E. Mueller, "Strategies for Minimizing Cancer Risk: Simple, Realistic Preventive Measures Could Save Hundreds of Thousands of Lives Every Year in Developed Countries Alone," *Scientific American*, 275, no. 3 (1996), 88–95.

affect a very specific outcome: cancer mortality. Thus, if your question is "How do we reduce cancer mortality?" it is fairly clear that cancer mortality is largely attributable to what epidemiologists call "environmental" causes—where the "environment" includes lifestyle factors like smoking, diet, and sedentary lifestyle.

Hereditary factors, or inherited mutations, appear to play a relatively small causal role in the vast majority of cancer deaths in the United States. At least a third to a half of all deaths due to cancer, especially cancers of the lung, breast, prostate, and colon, might be reduced or eliminated were we to eliminate smoking, poor diet, and other harmful lifestyle habits. The strongest evidence for this appears to be rates of death from lung cancer over the past six decades. The majority of women smokers started smoking a few decades after most men in the United States (many men received

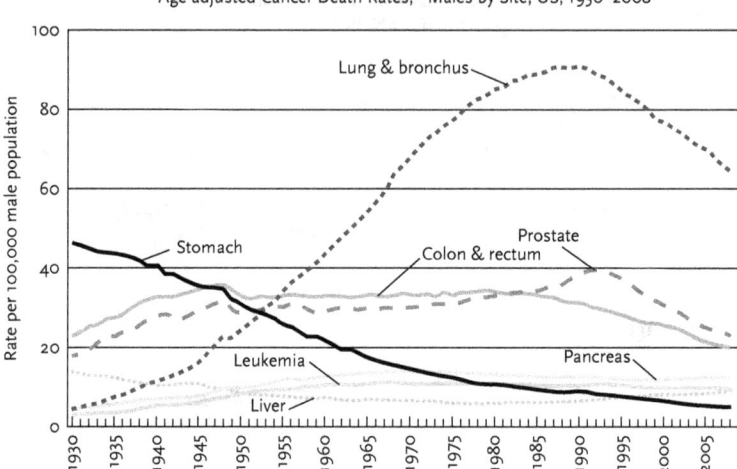

Age-adjusted Cancer Death Rates,* Males by Site, US, 1930–2008

*Per 100,000 age adjusted to the 2000 US standard population.

Note: Due to changes in ICD coding, numerator information has changed over time. Rates for cancer of the liver, lung and colon and rectum are affected by theses coding changes.

Source: US Mortality Volumes 1930 to 1959, US Mortality Data 1960 to 2008, National Center for Health Statistics Centers for Disease control and prevention.

FIGURE 3.4 Age-adjusted cancer mortality for US males, 1930–2008.

From a 2012 American Cancer Society report, "Cancer Facts and Figures" based on the CDC SEER database.

cigarettes in their rations during the world wars), and the curve of mortality associated with lung cancer follows the curve of start of smoking; it peaks much earlier for men and is still only gradually declining for women (Figures 3.4 and 3.5).

Despite these relatively uncontroversial epidemiological data (discussed further in Chapter 4), cancer researchers still say that cancer is a "genetic disease." What do they mean? Indeed, what could they possibly mean?

Talk of "genetic disease" appears to conflate at least two separate issues: heredity and proximate causation. The problem is not so simple as this, however, because the "same" genes (or mutations to those genes) can both play a proximate causal role in "sporadic" cancers and raise the chance of cancer in those bearing this hereditary mutation. That is, the same mutations to the same genes may be both acquired and inherited. There are many examples of "cancer genes" that play a role in both somatic and inherited forms of the disease, as might be expected for highly entrenched genes or biological entities with multiple functional roles. A vivid example is *BRCA1* and *BRCA2*. Inherited forms of variant alleles in *BRCA1* and *BRCA2* have been found to raise one's risk of breast and ovarian cancer. Women who inherit one of several variants of this mutation, and members of their family who share this variant, are more likely to develop breast or ovarian cancer than their peers; and they on average develop the disease at an earlier age. However, one does not need to *inherit* this

Age-adjusted Cancer Death Rates,* Females by Site, US, 1930–2008

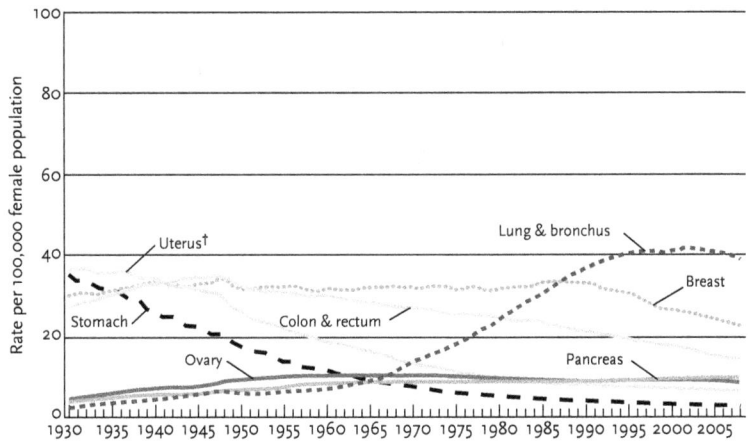

*Per 100,000, age adjusted to the 2000 US standard population. †Uterus cancer death rates are for uterine cervix and uterine corpus combined.

Note: Due to changes in ICD coding, numerator information has changed over time. Rates for cancer of the lung and bronchus, colon and rectum, and ovary are affected by these coding changes.

Source: US Mortality Volumes 1930 to 1959. US Mortality Data 1960 to 2008. National Center for Health Statistics, Centers for Disease Control and Prevention.

FIGURE 3.5 Age-adjusted cancer mortality for US females, 1930–2008.

From a 2012 American Cancer Society report, "Cancer Facts and Figures," based on the CDC SEER database.

mutation for it to play a causal role in the disease; one can also *acquire* it somatically. In sporadic cases of breast cancer, many cancer cells possess mutations to *BRCA1* and *BRCA2*. So the "breast cancer gene" is *both* a remote and a proximate cause of cancer.

That said, it is something of a mistake to speak of a "breast cancer gene." For one thing, because the gene (absent mutation) plays a key role in a variety of functions in the cell, the "breast cancer gene" is in fact a gene "for" DNA repair, among other functions. Moreover, there is no single mutation to this gene; it turns out that there are multiple variants of the *BRCA* mutation, with different *penetrance*. "Penetrance" refers to the proportion of individuals carrying a particular variant of a gene that also expresses an associated trait (or phenotype). Different mutations increase the risk of developing breast cancer to a greater or lesser extent. There are several estimates of breast and ovarian cancer penetrance in *BRCA1* and *BRCA2* mutation carriers. The estimated cumulative risks of breast cancer by age 70 years in two meta-analyses were 55–65% for *BRCA1* and 45–47% for *BRCA2* mutation carriers. Ovarian cancer risks were 39% for *BRCA1* and 11–17% for *BRCA2* mutation carriers (Table 3.1).

Table 3.1 Estimated Cumulative Breast and Ovarian Cancer Risks in *BRCA1* and *BRCA2* Mutation Carriers

Study	Breast cancer risk (%) by age 70 (95% CI)		Ovarian cancer risk (%) by age 70 (95% CI)	
	BRCA1	*BRCA2*	*BRCA1*	*BRCA2*
Antoniou et al. (2003)	65 (44–78)	45 (31–56)	39 (18–54)	11 (2.4–19)
Chen and Parmigiani (2007)	55 (50–59)	47 (42–51)	39 (34–45)	17 (13–21)

Note: CI = confidence interval.

From National Cancer Institute's Cancer Genetics Overview (https://www.cancer.gov/PublishedContent/MediaLinks/690485.html).

Inheriting mutations to *BRCA1* and *BRCA2* raises one's risk of breast cancer from the lifetime average risk through age 70 of about 12% to *anywhere* from 31 to 78%. That's a significant jump, but at the low end of the confidence interval, it is only double the risk of the average woman in the United States. The extent to which inheriting this particular form of the gene raises your risk over and above average lifetime risk may be relatively small; so inheriting this gene is by no means a death sentence. Perhaps not surprisingly, inherited mutations to this gene are more common, relatively speaking, than what are called, in contrast, high-penetrance "cancer syndromes" or "familial" forms of cancer. Such cancer "syndromes" are relatively rare: Li-Fraumeni syndrome, Lynch syndrome, familial retinoblastoma, adenomatous polyposis (APC), Von Hippel–Landau syndrome, and Werner syndrome occur in less than 50 of 100,000 individuals and play a role in less than 5–10% of cancers overall (NIH). In other words, establishing the relevance of this gene to cancer tells us very little (as yet) about its particular role or the value of relevant variables.

The vast majority of cancers are sporadic; they arise at later ages, and the mutations that play a causal role in these cancers are called "somatic" mutations. There are likely many germ-line (inherited) mutations that slightly elevate risk, even for sporadic cancers. That is, many common heritable genetic variations are associated with cancer susceptibility, but the vast majority have a lower penetrance than mutations associated with familial cancer syndromes. Most heritable mutations associated with cancer risk elevate this risk only slightly. Part of the reason that talk of cancer as a "genetic" disease is so confusing is that hereditary risk for cancer is not simply an "either/or" issue but a "more or less" one. The sense in which cancer is a "hereditary" disease is likely to become more and more a matter of degree as we learn more about hereditary dispositions associated with cancer. This is depicted in Figure 3.6.

In sum, the degree to which cancer risk is attributable to heredity varies, and high-risk mutations are in the minority, as might be expected from an evolutionary perspective. If a mutation accelerated the onset of disease as early as reproductive age or earlier, those bearing such a gene would not reproduce. Hereditary forms of cancer are relatively rare, and some may only double the baseline risk of disease. More generally, hereditary mutations that raise risk range in penetrance from exceptionally low to high, with most such mutations raising risk very little; we all probably bear some mutations that singly or in particular combinations (or combinations with other environmental factors) raise cancer risk.

Notice that the preceding discussion focused almost entirely on the problem of cancer risk; the question was whether and to what extent mutations to particular genes "accelerate" the onset of cancer or raise the risk of cancer. This kind of causal information is about how genes, or mutations to specific genes, are causally relevant to a population-level variable—cancer incidence. When scientists speak of a cancer as a "genetic" disease, they are often referring to this kind of causal information. But drawing upon Hausman's distinctions at the opening of

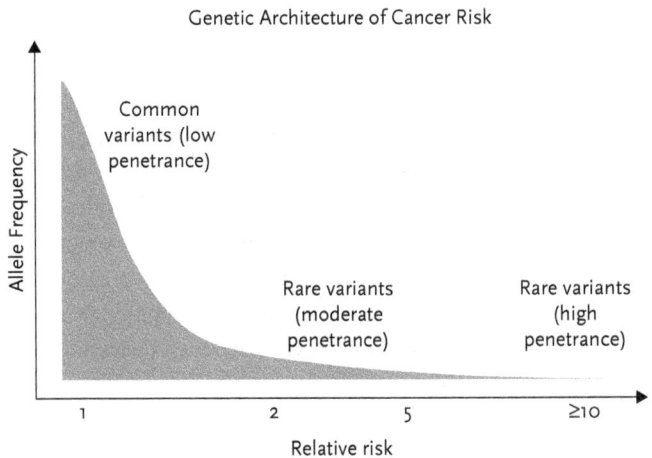

FIGURE 3.6 The relative causal roles of heritable mutations in cancer risk. Common, low-penetrance genetic variants are associated with low relative risk, and rare, high-penetrance variants are associated with high risk. There are few exceptionally high risk variants in the *BRCA1/BRCA2*. Mutations to these genes are associated with hereditary breast and ovarian cancer.

From National Cancer Institute, "Cancer Genetics Overview (PDQ°)—Health Professional Version" (https://www.cancer.gov/about-cancer/causes-prevention/genetics/overview-pdq), accessed January 1, 2017.

the chapter, we might have very different things in mind when speaking of cancer as a "genetic disease," and we should be careful not to confuse them:

- Mutations to specific genes are causally *relevant* to cancer incidence.
- Mutations to cancer genes play a causal *role* in cancer:
  - Where this causal role may be specific, as in this specific mutation to this gene, is *"the" actual difference-making cause* of some specific outcome at the cell or molecular level in some specific patient.
  - Or the causal role may be general, as in this specific mutation to this gene, whether inherited or acquired, will increase the chance of cancer, on average, in individuals.
- Mutations to a specific family of genes associated with cell repair and regulation of cell division may on average *raise the chance* of cancer in a population, particularly inherited forms. This is a sort of "type-level" claim—as in: we know that mutations of this type tend to increase the probability of cancer.

It seems relatively uncontroversial that mutations to some specific genes are relevant to cancer. However, the bar for causal relevance is low: almost everything is "genetic" (or environmental, for that matter) by this minimal criterion. The "parity" argument holds that different elements in the expression of biological traits have (at least in principle) causal parity (e.g., Griffiths and Knight, 1999; Maynard Smith, 2000; Oyama, 2000). More precisely, these authors emphasize that features picked out as distinctive of genes are not as distinctive as is often assumed. The call for "parity of reasoning" is, in fact, a call for consideration of factors outside the genome as playing a significant causal and explanatory role in biological traits.[6] For instance, epigenetic factors such as DNA methylation play a significant role in gene regulation and, in turn, in the cell cycle, as discussed in Section 3.3. Why then has there been such an excessive focus on genes?

We have stumbled here upon the "causal selection" problem, or perhaps better: problems. That is, there are a variety of ways to pick out causes as more or less significant. Philosophers often describe the causal selection problem as a single problem requiring a single answer. As described and explained by Hesslow, the causal selection problem is as follows:

> [A] normal event has many, perhaps infinitely many, causes, but . . . only some of them are selected and cited in causal explanations. Sometimes we even speak of the cause of an event which actually has several causes. Why

---

6. Thanks to Paul Griffiths for this clarification.

is this so? What determines the selection of the most important cause from the complete set of causal conditions? (1988, 12)

Hesslow's answer to this question is that there is no "metaphysical" difference between what we pick out as causes and background conditions; a cause is a cause. Indeed, most philosophers have agreed (see, e.g., Hart and Honore, 2002). There is no ontological difference between lighting the match and the presence of oxygen; both are causes of the fire or part of the "complete cause" or set of causal conditions necessary for the fire to come about. All grant that we as a matter of fact "select" some causes as more significant, and most take this to be for largely *pragmatic* reasons; which cause is explanatorily relevant depends upon the questioner and his or her interests and background assumptions. For instance, we all grant that oxygen is in most cases present, so we simply treat oxygen as a background condition we regard as fixed or stable and focus instead on causes we might be more likely to effectively intervene upon.

However, what the preceding brief discussion suggests is that the causal selection problem is not one but several different problems. When we ask about the "cause of X," the question is elliptical. For instance, we could mean "What are the main or predominant factors that increase the chance of X in populations largely similar to this population?" or, perhaps, "What are the proximate causes of this particular X?" or "What are the specific causes of this specific outcome, and how regularly are they associated?" These different requests for causal information call upon different kinds of evidence and, of course, might yield different answers. These fine distinctions are obscured when we consider the vague question "Is cancer a genetic disease?" or "Are mutations 'essential' to cancer?"

Cancer biologists might well argue that most cancer deaths might be prevented by the elimination of smoking or other environmental factors (the main or predominant causes of cancer mortality are environmental) and consistently claim that there is an important sense in which cancer is a "genetic" or "genomic disease," if they interpret this expression as simply referring to the fact that mutations to specific genes play a proximate causal role in the initiation and progression of cancer. This is not inconsistent, because the cancer biologist is shifting from talk of causal generalizations at the population level to talk of specific causal roles and causal mechanisms at the cell and molecular levels.

Moreover, each such causal claim is framed in terms of contrastives; or claims about causation always presuppose that there is some contrasting causal variable in play (cf. Hesslow, 1988). Which condition is singled out as "the" cause depends essentially on the composition of the reference class or contrasting causes we are considering. Once we specify the context in which a request for causal information is posed and the relevant contrast, there is a correct answer.

Waters (2007b) has defended the view that there is a meaningful "ontological" difference between the "actual" (or, some say, "specific") causes and merely "possible" causes. He takes himself to be resisting the mainstream philosophical view. Waters uses his argument to defend the view that genes are *causally specific difference makers* in a way that other causes of biological traits are not. While it may well be the case that for a given experimental population, we can (and do) legitimately pick out "actual" versus merely possible difference makers, as he claims, I disagree with his claim that scientists are interested primarily in "actual" difference makers. They are also interested in whether and how much they can generalize about causal relations across different populations; that is, they are interested in what Waters calls "possible" difference makers. They hope to discover how stable (and specific) such generalizations are and what background conditions make a difference. This is a different kind of causal selection problem, but still a very interesting and important one.

That is, scientists regard some causes as "background" and others as "specific" difference makers largely because they know the former to be relatively stable and the latter to vary. But stability is a matter of degree, and so too is "specificity." Indeed, the role of genes in cancer is an excellent case study in how both stability and specificity are a matter of degree. Let us turn to Waters's argument.

According to Waters, there is an ontological distinction between the merely "potential" and the "actual" difference makers:

> X is *the actual difference maker with respect to Y in population p* if and only if
>   i. X causes Y (in the sense of Woodward's manipulability theory).
>   ii. The value of Y actually varies among individuals in p.
>   iii. The relationship expressed by 'X causes Y' is invariant with respect to the variables that actually vary in p (over the spaces of values those variables actually take in p).
>   iv. Actual variation in the value of X *fully accounts for* the actual variation of Y values in population p (via the relationship X causes Y).
>   (Waters, 2007b, 567, emphasis added)

Are genes the "actual" difference makers in cancer in this sense? It depends upon the outcome of interest. With respect to specific gene products, mutations in genes like *p53* and *src* do qualify as the difference maker. However, with respect to "cancer" in general, no mutation to a gene is "the" difference maker for cancer, for actual variation in the presence or absence of a single mutations *does not "fully account"* for actual variation in Y, where Y

is the presence of a cancer. What is it to "fully account" for actual variation? According to Waters:

> Actual variation in the value of X **fully accounts** for the actual variation of Y values in population p (via the generalization "X causes Y") if and only if conditions (i)—(iii) above hold and (a) Individuals with the same X values in p have the same Y values. (b) An intervention on X with respect to Y that changed the X value of all individuals in p to the X value that one and the same individual had sans intervention would change Y values in p such that they no longer differed. (c) There is no variable Z, distinct from X, such that an intervention on Z with respect to Y that changed Z values in one or more individuals in p to the Z value that one of the individuals had sans intervention would change Y values in p. (Waters, 2007b, 567–568)

Mutations to a specific gene could not count as "the" actual difference maker for cancer because there are many "Zs"—the body is resilient and surprisingly resistant to cancer—and there are lots of ways in which other causal pathways intervene upon cancer cells to prevent progression to metastasis. So it is perhaps better to appeal here to Waters's notion of "an" actual difference maker when we are thinking of genes as causes of cancer. He explains as follows:

> In many biological situations, there is not just one actual difference maker, there are many. That is, actual variations in two or more variables in an actual population account for actual differences in an effect variable. In such situations, the operative concept is an actual difference maker, not the actual difference maker.
>
> X is an actual difference maker with respect to Y in population p if and only if
>
>   i. X causes Y.
>   ii. The value of Y actually varies among individuals in p.
>   iii. The relationship X causes Y is invariant over at least parts of the space(s) of values that other variables actually take in p. (In other words, it is invariant with respect to a portion of the combinations of values the variables actually take in p.)
>   iv. Actual variation in the value of X partially accounts for the actual variation of values in population p (via the relationship X causes Y).

Mutations to "oncogenes" and "tumor suppressor" genes are an actual cause of cancer in the sense that changes in the value of these variables are invariantly associated with the disease (at least in *part of the space* of values of other variables) in *some* populations. This is always a population-relative matter.

By way of illustration, consider the entire population of St. Louis. Suppose 33% of the women in St. Louis will be diagnosed with breast cancer in their lifetimes. Of that 33%, let's say that 2% may have a germ-line mutation to *BRCA 1* or *BRCA2*. If we consider the population of women in St. Louis as a whole, then a germ-line mutation to *BRCA1* or *BRCA2* is not "the" cause of cancer. However, suppose we narrow our considerations to the population of women with the germ-line mutation to *BRCA1* who in fact are diagnosed with breast cancer. For those women, the mutation would not even be an actual cause of cancer either, by Waters's lights, because the outcome in this population doesn't vary; that is, all the women have cancer. Suppose we were to reconfigure our population yet again and consider only those women with the germ-line mutation, some of which have the disease and some of which do not. Then the BRCA mutation becomes "a" cause (but not "the" cause, because other causes have to cooperate). This seems an odd result; that is, it seems odd that only in populations where the outcomes vary are mutations to BRCA I or II causes of cancer. Finally, consider yet another reconfigured population: of the 33% of women diagnosed with breast cancer, perhaps 5% will have a germ line to *BRCA1* or *BRCA2*. If we consider that 33%, it turns out that *BRCA1* or *BRCA2* is not even "a" specific cause of cancer in this case because, again, all the women have cancer. Yet surely it's true that the women who had the germ-line mutation were thereby more likely to get cancer. Genes seem to be either causes or not, depending on the population considered.

That is, depending upon the population we choose and the outcome at issue, whether a specific mutation counts as "the" actual cause, "an" actual cause, or merely a "possible" cause of cancer will vary. Waters anticipates this concern, a version of what he takes to be the standard "parity" objection in response to this observation: "The error of this possible objection is to infer from the fact that the selection of effect in an epistemic context involves pragmatics to the mistaken idea that what counts as the cause of that effect must also depend on pragmatics" (2007b, 19). In my view, this is a parody of the parity objection (cf. Griffiths et al., 2015). Waters notes that different ways of cutting up populations or different scales of analysis may in fact result in different identifications of genes as the cause (or no cause at all). For him, this does not matter, for this does not change the fact that for a specific population, there is a fact of the matter as to which cause is which. He can still maintain his thesis that there is an "ontological" difference between cause and condition or "actual" versus merely possible difference makers. Thus, in the population of myself and my brother, X may well be "the" actual cause of Y; but in the population of myself and my sister, X is not "the" actual cause of

Y. In his view, this is perfectly acceptable, because *actual causes are what biologists care about*: "Biologists are more interested in explaining actual differences than possible differences, and explaining actual changes over developmental and evolutionary time than possible differences. This is why they are more interested in identifying actual difference makers than possible difference makers" (2007b, 25–26).

I believe that this is a particularly misleading claim to make about both biology and biomedical science. Clinicians and researchers are interested in not simply *actual* causal differences; indeed, they care very *little* about actual causal differences in specific populations, such as the population of myself and my sister. Returning to Hausman, they are concerned not only that *X* is relevant to *Y* in some *actual particular circumstance*, but whether *X*s and *Y*s are functionally relevant to one another across a range of circumstances, how relevant, and which range of circumstances. Scientists care about "possible" and not simply "actual" causes because they know that in biology, uniformity of background conditions or functional organization is not to be expected. Particularly for complex diseases like cancer, the relevant permissive conditions are massively heterogeneous, and what may count as merely "possible" difference makers may turn out to be "actual" difference makers against different backgrounds in different populations. Cancer researchers would like to know not only why some individuals bearing these germ-line mutations do get cancer, but also why other individuals bearing the very same mutation do not, and to know this they have to consider a range of possibilities: perhaps they bear a protective mutation that halts the progress of disease. That is, they need to consider "possible" difference makers, which might ordinarily make little or no difference. Rare causal factors are still causal factors. Cancer scientists care about causal tendencies and causal relevance, not simply actual causal role, or even specific causal role—though, of course, these are useful pieces of knowledge to have (sometimes!). Indeed, a similar point is made by Griffiths et al. (2015), who defend an account of causal selection that relies on Shannon's measure of "information." On their account, causal selection is about reducing uncertainty; the more the uncertainty about a given outcome is reduced, the greater is the confidence we have in selecting a cause as significant. While I'm not entirely convinced that Shannon's information is a general measure that can address all the questions I have characterized as questions of "causal selection" (re specificity, stability, relevance, and role), it's clearly the case that scientists care about causes because they want to know how best to intervene. And they can better intervene if they can extrapolate across different contexts.

Waters could reply to this objection as follows: After all, these are causes that make an actual difference in those populations of individuals bearing the germ-line mutation, so why not call these causes actual difference makers as well in this population? This reply misses the larger point: what researchers studying cancer are coming to realize is that the assumption of uniformity of background conditions is deeply problematic. For instance, those same "protective" variations that may assist bearers of the germ-line mutation to *p53* could well be harmful in other genetic backgrounds. By identifying a gene or specific variant of a gene as "the" cause, we run the risk of making unwise extrapolations to other populations and other contexts. Moreover, populations evolve; what may be protective against one genetic background may not be protective against another genetic background three generations hence. Perhaps more importantly, genes are one cause along a series of causes in a progressive, degenerative process that affects and is affected by the immune system, sexual development, life history, and environmental exposures.

Recall that what distinguishes "hereditary" cancer syndromes is earlier onset—more than a single mutation is necessary before the disease develops. Actual variation in these germ-line mutated genes only partially accounts for actual variation in cancer's presence, in the sense that the presence of such mutations will increase the risk of cancer in most individuals possessing these mutations. "Oncogenes" are never (except in very restricted circumstances) "the" actual causes. The presence of any *specific* mutation is neither necessary nor sufficient for any specific cancer. While many cancers are strongly associated with specific mutations, these are not "necessary" in the sense that other mutations (and epigenetic events) are intersubstitutable chance-raisers for cancer, though not, of course, for exactly "the same" cancer.

Part of the challenge of multifactorial diseases like cancer is the contingency of causal relevance. That is, what arrives on the scene first determines what matters later, and even the order in which mutations occur can make a difference to the outcome. If we "hold fixed" or assume that one such mutation has already occurred, a specific downstream mutation may or may not be "a" cause for a given cancer in any given case; it may not be a "difference maker" for any specific cancer in a specific population. It all depends upon what else we hold fixed or what we put in our variable set and what population we happen to consider. In sum, it's not clear that Waters's account of actual causation is an effective way to approach multifactorial disease; we are interested in both common and diverging causal pathways, exactly because uniformity of background is not a given.

Clinicians care as much about "potential" causes as actual ones; they care as much if not more about the stability, or range of invariance of a cause, as its specificity (in fact, Woodward [2010] has anticipated and responded to this very concern). The more we learn about cancer, the more surprising it would be if two individuals shared exactly the same pathway to cancer: the exact same mutations, in the exact same order, to the exact same cells and tissues, followed by the exact same immune response and alteration in tissue microenvironment. While cancer researchers do seek and hope to find common pathways, they are as, if not more, interested in how and why pathways diverge.

As we saw in Section 3.3, what molecular biologists refer to as "mechanisms for cancer" are the entities, processes, and causal interactions that yield specific types of failure—failure associated with regulation of the cell cycle or cell birth and death, as well other capacities or hallmarks of cancer: the ability to attract a blood supply, metastasize, evade immune destruction, and reprogram the metabolism of the cell. Mechanisms for specific capacities are activated in context-sensitive fashion: they are local in their effects and highly unstable in terms of their consequences. For example, the same mutation (or epigenetic change, e.g., DNA hyper- or hypomethylation) can have different effects, depending not only upon which cell type and tissue of origin it occurs in, but also upon features of tissue architecture, sex or age of the patient, and even timing of onset. That is, the same mutation can have different effects on the progression to disease depending upon whether it arrives early (e.g., in the first dysplastic cells) or late (e.g., in the context of an advanced heterogeneous population of cells in a tumor).

3.5. Conclusion

So is there any sense to be made of talk of cancer as a "genetic" disease? One suggestion is that something is a genetic disease if and only if a certain proportion of individuals bearing some mutation are likely to acquire it. On this view, we can think of genetic disease as a continuum—some diseases surely do have strong hereditary components, and some diseases surely do depend for their development on changes to genes. One might think of "strong" genetic diseases as those where hereditary factors play a very significant role—as in Li-Fraumeni syndrome—or "weaker" genetic diseases as those involving hereditary factors that increase risk only.[7]

---

7. This view is defended by Smith (2007)—the "epidemiological account" of disease. He begins with two intuitions: what he calls the "individual" and "population" intuitions. First, if a disease is genetic, this must mean that those with the gene are more likely than not to develop the disease (or, more formally, if a disease is genetic, then $p(D|G) > 50\%$). Second, if a disease is genetic, cases of the disease more likely than not causally involve the gene (Smith, 2007, 97). Both of these intuitions drive Smith's account, which depends upon an analysis of disease risk at the population level. On his view, we should draw upon epidemiological concepts like attributable risk and population etiological fraction in considering diseases that may be classified as "genetic."

Attributable risk (AR) is an answer to the question "What percentage of the subpopulation having the gene will develop the disease in a way that causally involves that gene?" We calculate AR by dividing the total number of people whose disease causally involved the gene by the total number of people with the gene. If we know that AR is greater than 50%, then we know that most of those who have the gene will develop the disease. This addresses the "individual"

While this account of "genetic disease" may be one way to negotiate the problem of degrees of penetrance of inherited mutations that affect cancer risk, in general, my view is that talk of cancer as a "genetic disease" can mislead or confuse patients (and clinicians!) about the value of genetic information. After all, most cancers are sporadic, and there is an important sense in which it's ridiculous to speak of "genes for" cancer. After all, even *adaptive* genetic variation within the

---

intuition, described above. The population etiological fraction (PEF) is the number of individuals whose disease involved some cause X, divided by the total number of people with disease. This fraction is relevant to the intuitions Smith describes insofar as, if PEF is greater than 50%, then we have good reason to think that most cases of the disease in some population involve the gene. PEF pertains to the "population" intuition.

Smith claims that a disease is genetic whenever both PEF and AR are greater than 50%. He calls this a "minimally epidemiological" account of genetic disease, because it sets a threshold below which we would not (according to the intuitions he describes) regard a disease as genetic. The advantage of Smith's account is that it allows that "genetic" disease can be a continuum. At one extreme of genetic disease are cases where all those who have the gene develops the disease (provided they are not killed in a freak accident or die from another disease in their childhood), and the gene is causally involved in that disease's development. Li-Fraumeni syndrome would qualify in this case. Other, less strongly "heritable" cancers, like those associated with *BRCA1* or *BRCA2*, might be farther along the continuum toward sporadic cancers. To the extent that Smith's view captures the continuity of heritable contributions to cancer, it is helpful.

However, Smith's account has several disadvantages. The most obvious is his reliance on our "ordinary intuitions." There are, arguably, no "ordinary" intuitions about what would or should count as "genetic disease," and so his appeal to the "more often than not" criterion, drawing upon AR is, arguably, ad hoc. Moreover, the demand that both AR and PEF be greater than 50% seems arbitrary as well. However, even if we set aside these concerns, whether a disease is "genetic" on Smith's view may have to be age-indexed. All cancers develop over time (even "inherited" cancers). If one has acquired a clonal population of cancer cells, say, in one's colon by age 30, just for the sake of argument, all bearing mutations to $p53$, then one's chance of developing colon cancer by age 50 may be as high as 70%. But if one does not acquire this mutation until age 50, chances are that one will not live long enough to develop cancer. Arguably, there are many complex genetic factors that predispose one to developing earlier onset of cancer; these could be strongly heritable mutations to "cancer genes" or mutations to genes that affect development, or even taste and thus diet (a taste for sugar and carbohydrates may lead to a diet low in fiber and a higher risk for colon cancer). The "same" sporadic type of cancer would be both "genetic" and "not genetic," depending on the age of onset, as well as how strongly one's genetic profile predisposes one to various cancers, which will be no simple matter to discern.

While Smith's view is initially attractive in that it acknowledges that there are a continuum of cases where genes play greater or lesser roles in disease, one can surely dispute the intuitions upon which it relies. It seems ad hoc to demand that one must be more likely than not to develop the disease when possessing the gene. One could, for instance, require that the probability of developing the disease, given a gene, is simply greater than the probability of developing the disease without the gene. Further, the account rests upon a notion of attributable risk, which, unfortunately, does not do the work Smith hopes for it.

Attributable risks can, unfortunately, sum to more than 100. For instance, in one study, sudden infant death (SID) might be attributable to prone sleeping versus back sleeping in infants in 60% of all cases; in another study, SID might be attributable to parents smoking versus not smoking in 70% of all cases. Clearly, both factors are causally significant and relevant

normal range may contribute to cancer. As mentioned previously, many cancers of the breast and prostate are sensitive to estrogen and testosterone, hormones that enable healthy sexual development and reproductive success. These hormones are presumably under the control (indirectly) of many genes. Are these "genes for" cancer? Perhaps genetic variants that contribute to slightly elevated levels of such hormones might be said to be a "risk factor" for such cancers, but genetic variants associated with elevated hormone levels are surely not "genes for" cancer. Or perhaps it is better to say that there are no genes "for" cancer, only genes, mutations to which play some causal role or other in cancer.

Oncogenesis is a case study in complex causation. As such, it presents a challenge and opportunity for philosophers of science to test many of our beloved theories about causation, evidence, explanation, and confirmation. Good evidence for causal claims in molecular genetics and genomics is rather different from the kinds of evidence sought and discovered in epidemiology. This is at least in part because the objects of explanation, and the temporal and spatial scale at which causal claims are asserted and confirmed, are significantly different. In part also, it is because we are concerned with different kinds of causal claim in such different disciplines: claims about causal relevance versus causal role, types versus tokens, or irregular causal generalizations, versus claims about specific values for specific variables. So, while some general principles of what count as "causes of cancer" can be defended across the disciplinary contexts, the causal reasoning at work in different disciplines is rather different. Moreover, some talk of cancer as a "genetic" or "environmental" disease is rhetorical, and practical, or informed by what we believe we can intervene upon more or less successfully. This lends credence to the idea that, in practice at least, judgments about causation are in some sense "normative" or closely tied to judgments about what we believe we

---

to crib death, but the numerical gloss on AR is deceiving. Attributable risk is non-additive; all that AR measures is the significance of the presence or absence of some exposure in some given population, not the relative significance of this exposure versus other causal factors. That is, AR measures the relative likelihood (in some specified sample) of developing the disease, given some risk factor versus the lack of that risk factor. This measure is always relative to one study and some study context. This measure will, of course, vary with one's sample; this is simply Lewontin's point in his heralded "Analysis of Variance and Analysis of Causes" paper. Attributable risk for genetic contribution to disease in one population (say, 70%) may well be identical for environmental contribution in a different population (say, 70%).

Smith grants the population relativity problem, but he seems to place it offscreen. That is, he acknowledges the difficulty of attempting to *discover*, in general, for any population, what role genes versus environment play in some disease, given the problem of extrapolation, but simply argues that if *it were possible to discover this "absolute" number*, we *could* attribute some disease to genes on the grounds he provides. But this is to ignore that we simply cannot assume or hope that the population(s) we've sampled are all homogeneous with respect to all possible confounding causes or that the gene is the major or only significant cause on the basis of AR alone.

may most successfully intervene in (see, e.g., Hitchcock and Knobe, 2013). I have argued that the mechanistic research program in cancer has been fruitful, but mechanistic understanding of the role of mutations in the etiology of cancer is only part of the picture; we need to zoom out on populations and their history to better understand disease etiology, or so I will argue in the next chapter, on environmental causes of cancer, and the chapter following, on cancer as a product and process of evolution.

4 EVIDENCE AND ENVIRONMENTAL EPIDEMIOLOGY

A PRAGMATIC APPROACH

All scientific work is incomplete—whether it be observational or experimental.
All scientific work is liable to be upset or modified by advancing knowledge.
That does not confer upon us a freedom to ignore the knowledge we already
have, or to postpone the action that it appears to demand at a given time.
(Hill, 1965, 12)

4.1. Introduction

Sir Bradford Hill, the author of the above passage, knew of what he
spoke. He was one of the first epidemiologists to describe an association between smoking and lung cancer, and his research was not (initially, at least) embraced with open arms. As Hill knew all too well, it
is extremely difficult to gather evidence in epidemiology. Particularly
where cancer is the outcome of interest, identifying associations between exposures and outcomes, and ruling out confounders, is extremely difficult. Because cancer takes decades to develop (by and
large), research on environmental causes of cancer is time-consuming
and expensive, often drawing upon multiple lines of evidence from
different disciplines with different methodologies (epidemiology,
toxicology, genetics, molecular biology, or clinical medicine). Causal
inference in such cases is a process that often begins with partial, incomplete, and tentative hypotheses, perhaps suggested by population-
level associations, which lead to "established" risk factors only after
decades of debate over potentially confounding causes. Of course,
epidemiologists' arguments can never fully establish that X is a cause
of Y, if what we want is perfect certainty; but, arguably, nothing can.
Science is, after all, a fallible enterprise. Thus, asserting that some association is causal—even in cases where all the evidence is in favor—
involves taking an epistemic risk.

Nonetheless, for several decades now, epidemiologists have been fairly confident that they can and should take such a risk. Epidemiologists by and large have agreed that environmental factors are implicated in the vast majority of cancer deaths (Doll and Peto, 1981; Perera, 1995; Moeller, 2009). In other words, cancer is considered an "environmental" disease in the sense that it's now widely held that intervening on a variety of exogenous factors—like tobacco use or occupational hazards—would reduce mortality from cancer significantly. According to the National Cancer Institute, about 30% of cancer deaths are attributable to tobacco use, and as much as an additional 30% of cancers might be prevented by changes in diet, vaccinations for infectious disease (human papillomavirus, hepatitis C), reduction of exposure to ionizing radiation, and reduction of occupational risk factors (NCI, 2012). References to cancer as an "environmental" disease, or claims to the effect that cancer risk is "attributable" to environmental causes, are fairly typical. However, such claims are often misunderstood, for two reasons. First, "environmental" causes in the popular imagination are limited to causes outside of one's control (air pollution, water pollution, etc.). Environmental epidemiologists, in contrast, take "environmental" causes to include any causative factor external to heredity, whether ingested, inhaled, or absorbed, whether a matter of choice or not (Moeller, 2009).

Second, and more seriously, talk of "environmental risk factors" and terms like "implicated" and "attributable" are vague, suggesting something less than cause but something more than association. In environmental epidemiology of cancer, "association," "attributable risk," and "risk factor" are more commonly used than the far more contentious term "cause." In part, this is because the former terms have more precise meanings. For instance, "attributable risk" measures the difference in risk of outcome between exposed versus unexposed populations. Many epidemiologists are extremely reluctant to use attributable risk as a proxy for cause, perhaps not surprisingly, given the long history of legal and political battles about the interpretation of epidemiological evidence for causal associations, for example, between smoking and lung cancer (Hill, 1965; Rothman, 1976; Rothman and Greenland, 2005) or between the pesticide DDT and the health of humans and wildlife (Carson, 1962; Dunlap, 1975, 2008). Indeed, there is a long-standing debate among epidemiologists (and, more recently, among philosophers) about when epidemiological evidence can provide sufficient grounds for causal claims, as opposed to more precise and circumscribed claims about attributable risk (Broadbent, 2013).

There are two extremes along a continuum of views. At one extreme, some epidemiologists refuse even to speak of causal relationships, preferring to speak only of more precisely defined measures of attributable risk or relative risk. Similarly, some philosophers have adopted the view that the evidence provided by epidemiological research, such as case control studies, is not sufficient warrant for causal claims. For instance, Russo and Williamson argue: "To establish causal claims, scientists need the mutual support of mechanisms and dependencies. The idea is that probabilistic evidence needs to be accounted for by an underlying mechanism before a causal claim can be established" (2007, 159). There are a variety of more or less critical and conciliatory responses to this claim in the philosophical literature, known as the Russo-Williamson Thesis, or RWT (Weber, 2009; Broadbent, 2011c; Gillies, 2011; Leuridan and Weber, 2011; Howick, 2011; Illari, 2011; Reiss, 2015). Those critical of the RWT are willing to grant that epidemiological research is perfectly adequate to warrant causal claims, provided various conditions are in place. Of course, there is some debate over what these conditions are; indeed, this has been the subject of an ongoing debate among epidemiologists for decades (Hill, 1965; Rothman, 1976; Rothman and Greenland, 1998, 2005; Skrabanek, 1992, 1994; Vandenbroucke, 1988; Weed, 1986; Morabia, 2004; Broadbent, 2013). To some extent, many of these debates echo some of the very same debates concerning genetic causation discussed in Chapter 3; critics of competing views often talk past each other because of their failure to specify whether what is at issue is mere causal relevance, role, or specificity.

One of the most well-known statements on this topic is Hill's 1965 paper, "The Environment and Disease: Association or Causation?" Hill's paper was (in part) an attempt to address some of the concerns that had been raised with respect to his and others' studies. Hill lists a variety of considerations relevant to claiming that an environmental variable is causally responsible for a given disease; among them are strength of association, consistency across different backgrounds, specificity of causal role, temporal priority, dosage effects, or "biological gradient," and, last but not least, plausibility of underlying biological relationship. Hill's "considerations" are *not intended to be exhaustive criteria* for claiming that X is a cause of some outcome (i.e., they are not necessary and sufficient conditions). Rather, his paper is best read as a plea for balanced consideration of the total available evidence. It is also an appeal to precaution, especially where harms are significant. In other words, Hill's paper advances both an argument about evidence and a *normative* argument for a legitimate role for evaluative judgments in matters of public health.

The goal of this chapter is in large part to defend Hill's views, drawing upon two case studies in environmental epidemiology: the case of smoking and lung cancer and the case of the Downwinders, or the people in Utah and southern Nevada exposed to radiation fallout after nuclear testing in the 1950s. These are two extremes along a continuum of cases that meet Hill's conditions to a greater (smoking) or lesser (Downwinders) extent. Whether we take the evidence to suffice will depend upon what we use this evidence in service of. Riess (2015) has defended what he calls a "pragmatic" theory of evidence, which also draws upon Hill's account, and I take the current chapter to be very much in the same spirit.

It should be relatively uncontroversial that evidential warrant is a matter of degree, and epidemiological evidence can provide greater or lesser support for causal claims. Hill offers a variety of pragmatic considerations of relevance to assessing the strength of warrant for causal claims: strength of association, consideration of and elimination of relevant confounders or contrasting causal explanations, dosage effects, identification of likely mechanistic bases, and so on. Reiss (2015) argues for elimination of confounders as essential. In my view, no one of these is necessary or sufficient on its own; rather, considerations of warrant need to balance all of these considerations. Moreover, when and whether we have "sufficient" evidence depends ultimately, in my view, on the question "sufficient for what"? Given that evidence for causal claims may be weaker or stronger, we may have degrees of belief in that warrant action—whether regulatory action, or compensation for harm, as in toxic torts—in some contexts but may well demand stronger evidence for others.

Royall (1997) distinguishes three questions about evidence that may help make this point clearer:

- What does the evidence say?
- What should I believe?
- What should I do?

As Royall explains, describing what the evidence "is" is not the same as telling us what to believe. One can well have ample evidence of one kind or another and not believe that an association is causal. Further, whatever the evidence says, and even whatever one believes, with whatever degree of confidence, are (relatively) independent of what one should do. Suppose one believes one has offended a colleague, but is far from certain. It may be wise to apologize and thus avoid further potential conflict in advance of collecting further evidence, all things considered. Different questions about what to do in different circumstances may well require different evidence. So what it is to have "sufficient" evidence must depend on how the evidence is put to use. In many cases where "sufficiency" is at issue, we are debating whether to act. The question at issue is not "Do we know this causes that?" but "Do we have enough evidence to warrant this or that action?" This is particularly vivid in many cases of debates over environmental causes of cancer.

The demand for "mechanistic" evidence in the philosophical literature obscures the way(s) in which debates about evidence, belief, and action are intertwined. Given that the time between exposure and outcome in environmental causes of cancer can span decades and that the causes are multiple, demanding evidence of "mechanisms" in environmental causes of cancer is not only vague, but dangerously so. It leaves open to skeptics a possibility of endlessly contesting evidence for causation that is as decisive as may be reasonable for establishing public policy. As is evidenced by a long history of acrimonious debate over environmental causes of cancer (Oreskes and Conway, 2010; Proctor, 1996), such demands can be disingenuous attempts to defray the expense of regulation or deflect responsibility. In contexts where inaction can cost significant loss of life, as Hill argued, we ought to consider the total evidence (e.g., strong association or association established across a wide variety of contexts), and whether or not this evidence is decisive once and for all, it may well be more than sufficient reason for taking precautionary measures.

Even on a weak construal of the RWT, we arrive at a fairly ambiguous demand (cf. Illari, 2011). What it is to identify or provide evidence for a mechanism linking cause and effect varies across different disciplinary contexts and over time. As Illari (2011) has argued, even the concept of "mechanistic" evidence is multiply ambiguous. There are many different types of evidence for mechanisms, ranging from detailed accounts of the particular biochemical entities and processes linking a given cause and effect, to evidence that there is a mechanism of some type or other, to evidence about similar or analogous mechanisms of the same type, for example in model organisms. There is even evidence of absence—that is, a good grasp of what plausible confounders might be and an argument that they may be ruled out (on the basis of a case-control study, for instance, that matches subjects with and without the confounder). There are many kinds of evidence that plausibly answer a demand for "mechanistic" evidence.

In sum, different contexts might reasonably require different standards, exactly because we may have good reason to act, even before we have anything like a "full" accounting of the mechanisms in operation (whatever that might mean). The reverse is also true; sometimes we want to await a full accounting, exactly because action will be so very costly. That is, there are *pragmatic* and context-dependent reasons why we should either wait for better evidence or not. When translating science into policy domains, we ought to consider demands for evidence as serving a variety of functional roles; the demands we place on the quality and character of the evidence depend essentially on what the evidence will be used for. For instance, we might well invoke different standards in some policy contexts versus others, or in regulatory versus tort law, or in public recommendations from national bodies versus contexts in which individual patients consider a course of action with their clinician. When assessing the variety and quality of evidence, we need to take into consideration the costs of failing to act. Where salmonella-tainted hamburgers are killing children, good epidemiological detective work may require evidential reasoning that looks far less like the purported gold standard—or highest standard—of a randomized controlled trial (RCT).

What we need the evidence for should help us decide what kind of study is necessary or how to gather and assess evidence in any particular case. How "deep" or "wide" the evidence ought to be (and whether such evidence must include a "mechanism") must depend on how the evidence is used. Several philosophers of science have argued that values are an essential component of evidential reflection, for just such reasons (Douglas, 2000; Brown, 2014).

The case studies of environmental cancer causation discussed in this chapter range from clear-cut to contentious. The aim is to gain some sense of the process of reasoning about causation in epidemiological research and the *function* of demand(s) for evidence, as well as the function of various tools that have been used to challenge evidence or mislead us about such evidence. That is, sometimes the function of demands for more evidence is simple obfuscation; industry interests, for instance, spent decades demanding further "proof" of a link between tobacco and lung cancer (Oreskes and Conway, 2010). And in legal or policy contexts, defendants in toxic torts cases frequently demand ever more detailed mechanistic evidence as a way to induce skepticism in judges and juries. In these cases, demands for more mechanistic evidence serve the aim of contesting precautionary policies that require costly changes to practice (Cranor, 2011). It's fairly clear in many such cases that the function of demanding evidence is politically or financially motivated. However, sometimes demands for more detail are well motivated; we might have good empirical reasons to suspect confounders or common causes. Knowing which confounders are likely depends upon background knowledge and specialized expertise in disciplines as diverse as genetics, developmental biology, and chemistry. This is exactly why inference in the context of public health is an iterated, multidisciplinary enterprise and should not be left to judges or juries to assess (see, e.g., Cranor (2006)).

A variety of organizations, such as the National Institute of Environmental Health Sciences (NIEHS), the US National Toxicology Program, (NTC), the International Atomic Energy Agency (IAEA), and the International Agency for Research on Cancer (IARC), have official methods of assessment, and classifications for carcinogenicity based on standardized criteria. The quality of evidence is assessed in light of the nature and extent of epidemiological and animal (toxicological) data, as well as experimental work on the basic biochemical mechanisms involved. These standards are regularly updated on the basis of new kinds of evidence or new ways of combining or assessing evidence (IARC Preamble, 2006). Determination of carcinogenicity by national or international bodies like the IARC is a consensus procedure, and so is in part scientific but also in part a political process. In other words, what makes something count as a carcinogen in the view of the IARC is socially negotiated rather than a strictly empirical matter (though, of course, all science is in some sense a matter of social negotiation). Thus, understanding why and whether something is officially considered carcinogenic would require a very different investigation than this one: focusing on the social epistemology governing these decisions, the history and political forces shaping institutions like the IARC (for a discussion, see Leuridian and Weber, 2011). This chapter focuses primarily on the science, not because these issues are not socially negotiated, but because the aim is to clarify what science is relevant and how in determining carcinogenicity, not how this is decided by government or international bodies.

4.2. Evidence in Epidemiology: An Overview

Classical analytic epidemiology is the study of associations between some exposure and outcome, and the field is concerned primarily with interventions on such exposures, where relevant, in service of public health. There are various types of epidemiological studies. Sometimes epidemiologists rank the strength of evidence in establishing an association in terms of the relative closeness of their results to the "ideal," namely, the results of a randomized controlled trial: "Other things being equal, results from randomized trials, can offer a more solid basis for an inference of causes and effect than results obtained from any other study design" (Koepsell and Weiss, 2003, 98). Reiss (2015) characterizes this as the evidence-based medicine, or "EBM paradigm" of evidence, a model that takes the RCT as the gold standard for scientific research in medicine and public health (for a critical history, see, e.g., Solomon, 2015). The epidemiological "design tree"—a map of different kinds of epidemiological studies—ranks such studies as closer to or farther away from this ideal (see Figure 4.1).

A non-randomized or "observational" study, we are told, is subject to confounding: "distortion of the exposure–outcome association due to their mutual association with another factor" (100), or what philosophers call "common causes." Because subjects are not randomly assigned to exposures—smoking, genetics, lifestyle choice, physical or social environment—there will be systematic differences between exposed and non-exposed populations that could serve as common causes of some measured effect. Thus, some epidemiologists rank "quasi-experimental" studies (where exposures are assigned, but not at random) as less than ideal, followed by cohort studies (either prospective, i.e. following groups over time and tracking exposures and outcomes, or retrospective), case-control, cross-sectional, longitudinal, and "ecological" studies, which they rank as farthest away from the ideal.

As an example, a prospective cohort study was conducted from 1968 through 1988 on women's contraceptive use and the occurrence of cancers of the reproductive tract. Half of the women were users of oral contraceptives at the time of their recruitment, and half were never-users. Beral et al. (1988) found that oral

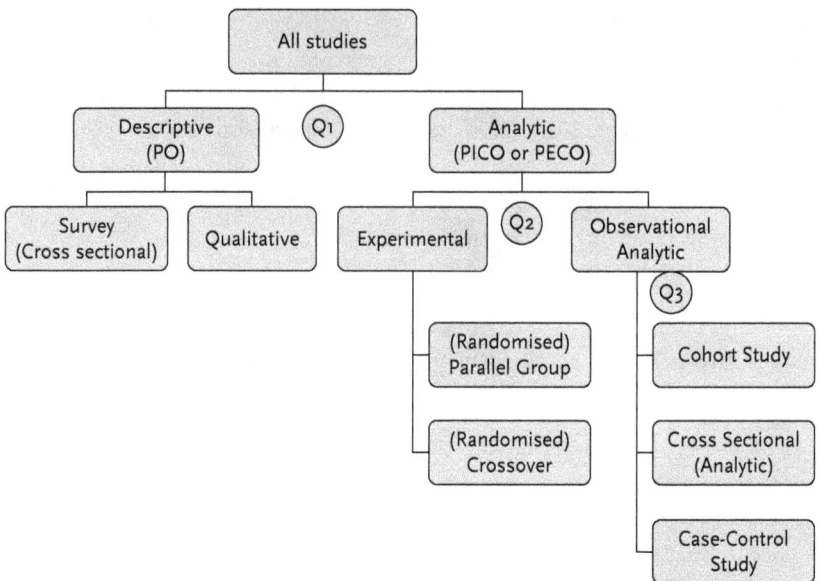

FIGURE 4.1 The image represents the variety of types of epidemiological study design. A descriptive study characterizes the prevalence, incidence, or experience of a group. Examples of such studies include case reports, or surveys. In contrast, an analytic study attempts to quantify the relationship between two factors – an intervention or exposure, and some outcome. In experimental studies, an experimenter intervenes on the intervention variable. In contrast, in analytic observational studies (the vast majority of epidemiological research), researchers measure the exposure or "case" against control groups. Center for Evidence-Based Medicine (http://www.cebm.net/blog/2014/04/03/study-designs/), accessed January 1, 2017.

contraceptive use was associated with a two to three times higher incidence of cervical cancer and a lower incidence of uterine and ovarian cancers. Despite the fact that this association could be explained by some common cause (say, greater exposure to sexually transmitted diseases among users of oral contraceptives, yielding higher rates of cervical cancers), this study is considered closer to the ideal, in establishing causation, than other kinds of epidemiological studies.

For instance, a longitudinal study simply tracks some suspected causative factor over time, mapped against some outcome measure, within or across different populations. Apart from confounding, this kind of study is subject to one of the notorious flaws of epidemiological studies. The "ecological fallacy" is when a population-level association is erroneously taken to imply an individual-level association. That is, just because on average in a population, for instance, X correlates with Y, this does not mean that any individual characterized by X is more likely to be Y. That smoking is (generally) a greater risk factor than others for death from lung cancer does not mean that Mr. Smith (who is dying of lung cancer) was a smoker. Despite their limitations, ecological studies can be informative or, at minimum, suggest hypotheses for further testing. For instance, Figure 4.2 shows the results of an ecological study that examined rates of cervical cancer and screening practices in Finland, France, and Romania. Rates of cervical cancer were significantly higher in nations with less well-organized or systematic screening practices.

This study at least suggests that implementing national Pap screening can have a significant impact on rates of cervical cancer. For defenders of evidence-based medicine, this kind of study falls short of the ideal of randomized controlled

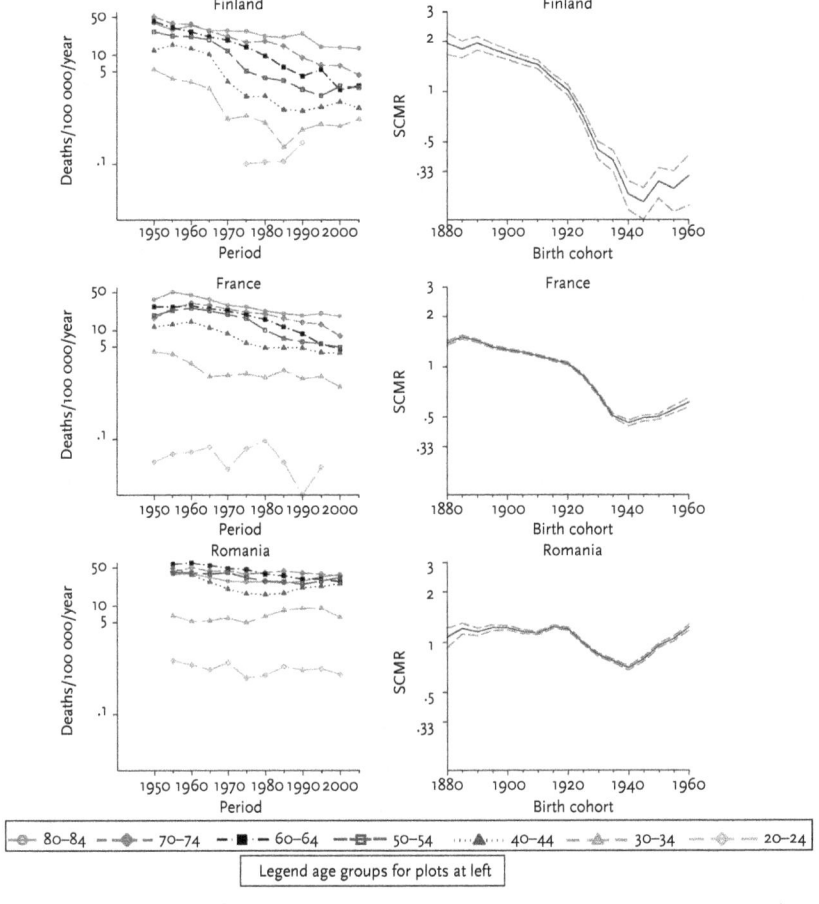

FIGURE 4.2 Results of a descriptive study of cervical cancer incidence over time in three countries (Finland, France, and Romania) that instituted different Pap screening policies. From Marc Arbyn, Amidu O. Raifu, Elisabete Weiderpass, Freddie Bray, and Ahti Anttila, "Trends of Cervical Cancer Mortality in the Member States of the European Union," *European Journal of Cancer*, 45, no. 15 (2009), 2640–2648.

trials, where (at least in principle) all possible confounding factors are controlled for. While some epidemiologists often seem to take randomization as the pinnacle toward which any epidemiological study should strive, philosophers are far less sanguine about the merits of the RCT. Several philosophers (Worrall, 2002, 2007, 2011; Solomon, 2011; Cartwright, 2011) have raised a number of objections to RCTs. But with respect to the question of standards of evidence in epidemiology, a larger point deserves mentioning. Any empirical study of causal relationships in populations is going to fall short of "proof." As Worrall comments in passing, "A positive result in an RCT (randomized clinical trial) does not establish causality—nothing can" (2011, 236).

Impossible standards aside, several epidemiologists have lately adopted one or another statistical approach in order to better approximate an ideal method for establishing causation (e.g., path analysis, Bayesian nets). While such approaches provide a model for how we might discover causes *in the ideal case*, they are often difficult to use as a way of adjudicating actual causation evidence that is far from the ideal, which is most of the time. It is easy enough to say that causation obtains if and only if when one (at least in principle) manipulates the value of some variable, x, the value of some second variable, y, changes. It is quite another to announce that one has discovered a causal relationship when no such manipulation is possible, either for pragmatic or for ethical reasons. This is overwhelmingly the case in the context of environmental cancer causation.

How, then, do epidemiologists provide evidence for environmental risk factors in cancer? The short answer is that they piece together many independent bits of evidence—from blood or tissue sample data with biomarkers of exposure or from case-control, cohort, or ecological studies—and link this with both data on mechanisms and experimental work in toxicology. In essence, this is a process of inference to the best explanation (cf. Broadbent, 2013). Many philosophers view this as an extremely tenuous method of confirmation. However, as Peirce saw it, abduction was not a method of confirmation or discovery, but in fact both. Peirce lived in an era before philosophers came to regard the distinction between the two as sharp, and in fact, if we look to scientific practice, inference to the best explanation (IBE) is not a matter of either confirmation or discovery, but often both. Scientists engaged in establishing a link between some environmental variable and a change in cancer incidence and mortality are often engaged in iterated processes of the formulation of plausible hypotheses that "would" explain, intertwined with showing how and why such hypotheses most likely "do" explain. In the domain of public health, if several independent pieces of evidence all point in the direction of a causal association, one *at least* has grounds for exercising precaution and monitoring some suspected cause of cancer.

I now turn to studies that range from what we take to be uncontroversial, well-established cases to what some might consider speculative. Apart from the fraught political context of many of these cases, the three challenges I use to frame the discussion are based on Royall's (1997) three questions, as listed above: What does the evidence say? What should we believe based on the evidence? And what should we do we do? Royall's questions serve as a reminder both that these questions can be addressed independently and that questions about evidence cannot be reduced to questions about warranted belief, without losing sight of a great deal of context of relevance to both scientific practice, and the practical import of science.

4.3. Tobacco

There are many known or suspected environmental carcinogens, but by far the most well established is tobacco. Others include air pollution, particularly that containing particulate matter; byproducts of fire retardants, toys, cosmetics, and other plastics products; organochloride pesticides and fungicides; and drinking water disinfection byproducts. Causation is much more difficult to establish in the case of such diffuse, low-level exposures. Exactly because individuals do not choose when and where and how much they are exposed to such chemicals, establishing a causal link requires many independent pieces of evidence, both "top-down" (e.g., using biomarkers to establish associations between disease and exposure) and "bottom-up" (molecular mechanisms of action). In contrast, one can see the evidence of smoking, literally, in the tissue of a smoker's lungs.

To ask Royall's first question, what is the evidence? The epidemiological evidence for the link between smoking and lung cancer was first collected in the 1940s and (most famously) in the 1950s, with the work of Wydner, Doll, and Hill. In defiance of the beliefs of most mainstream physicians, these three epidemiologists doggedly gathered evidence linking smoking and the marked increase in lung cancer in the first decades of the twentieth century. Needless to say, some controversy ensued, and there was a good deal of pressure, particularly from the tobacco industry, to challenge the data and the conclusiveness of this evidence (Oreskes and Conway, 2010). This history has been amply covered by both popular and academic histories, so I will confine my summary to the most relevant scientific milestones.

Some of the most well-noted publications establishing the link between smoking and lung cancer appeared in the 1950s. However, there was research linking smoking and lung cancer as early as 1939. In Germany, Mueller (1939) found that a majority of surveyed lung cancer patients were regular smokers, whereas only three of eighty-six such patients he surveyed were non-smokers. However, the sample sizes of Mueller's and other studies done in the United States in the 1940s were small. It was not until 1950 that populations of significant size were surveyed, notably by Wynder and Graham in their landmark "Tobacco Smoking as a Possible Etiologic Factor in Bronchiogenic Carcinoma: A Study of 684 Proved Cases," published in May 1950. Shortly thereafter, in September of the same year, Doll and Hill published "Smoking and Carcinoma of the Lung," based on a population sample in the UK. The studies were similar in some respects but rather different in others. First, both were attempting to explain the same observation: an increase in the incidence and mortality from lung cancer in the preceding half century. However, Wynder and Graham focused on the increase in incidence, whereas Doll and Hill were concerned with mortality. Second, both drew upon rather large samples: 684 cases in Wynder and Graham and 709 in Doll and Hill. Wynder and Graham found that of 605 male patients with proven bronchiogenic carcinoma of the lung, 1.3% were non-smokers, whereas 51.2% of them had smoked twenty or more cigarettes a day over the previous twenty years. Fully 95% of the 605 patients were moderate or heavy smokers, as compared with 73.7% of non–lung cancer patients in the same sex and age cohort in the hospital population. Fully 96% of the lung cancer patients had been smokers for over twenty years, and 51.2% of lung cancer patients were excessive or heavy smokers, as compared with 19.1% of the general hospital population. In other words, lung cancer risk seemed to increase in proportion to the duration and extent of smoking habits; lung cancer was exceptional or rare in non-smokers. Data were obtained primarily during personal interviews (634 cases), or where this was not possible (about thirty-three cases), either questionnaires were mailed to the participants, or (17 cases) interviews were conducted with persons "intimately acquainted with the patient throughout his adult life." (At least 17 patients had died by the time the interviews were conducted.) Interviewers assessed the patients' smoking history, job history, exposure through occupation to dust, fumes, or insecticides, age, alcohol use, birthplace and geographical location, cause of death of first-degree family members, and site of lesion, as well as the method of diagnosis. Similarly, patients in the control group were matched with the cases by age and interviewed about their smoking habits, occupation, and so on. The controls were either "general hospital patients" with "diseases of the chest that were not lung cancer" (control group I) or patients in various hospitals around the country (New York, Cleveland, Illinois) being treated for other ailments (control group II).

Like Wynder and Graham, Doll and Hill used a case-control method; 709 individuals surveyed were controls—individuals with diseases other than cancer—709 were patients with cancer of the lung. The study began in 1947, when a group of London hospitals were instructed to ask patients diagnosed with cancer to take part in an interview with an "almoner"— a person dispatched to conduct interviews with patients and randomly selected controls at the same hospital. Both cases and controls were interviewed about their smoking habits—duration, amount smoked daily, occupation, place of residence, and "social class" (presumably determined by income). Doll and Hill found that the risk of lung cancer increased with both duration and extent of smoking. They also included a graph in their paper showing the death rate from lung cancer on the x axis and the consumption of tobacco per pound per person, from 1900 through 1950, on the y axis (Figure 4.3). As they remarked, "In the quarter of a century between 1922 and 1947 the annual number of deaths recorded increased from 612 to 9,287, or roughly fifteen fold. This remarkable increase is, of course, out of all proportion to the increase of population—both in total and, particularly, in its older age groups" (1950, 4683).

In other words, Doll and Hill explicitly framed their research with the data on the dramatic rise in lung cancer deaths since the early part of the Century (not

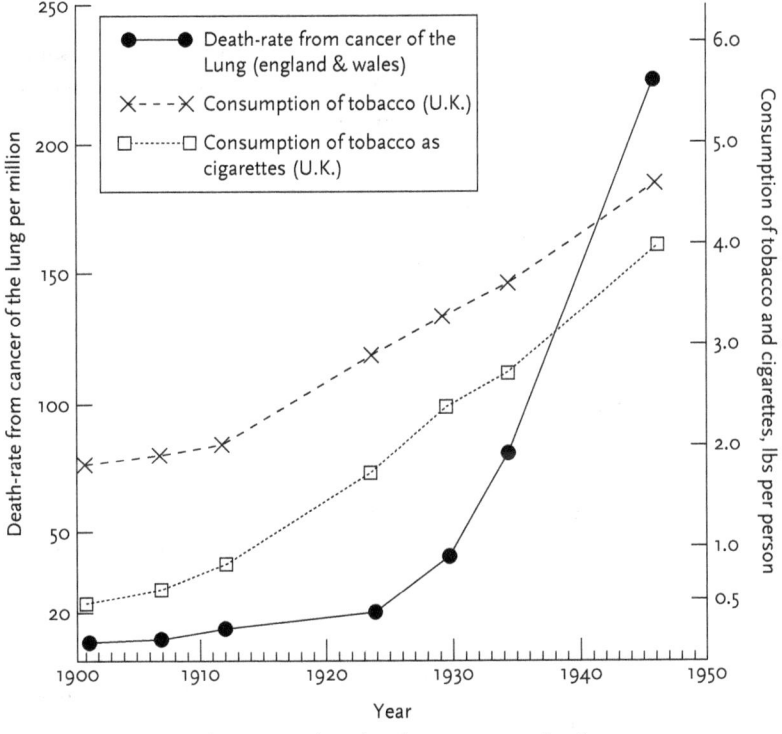

FIGURE 4.3 The iconic image from Doll and Hill's study of smoking and lung cancer. This shows patterns of cigarette consumption from 1900 to 1947 and the parallel increasing incidence of death from cancer of the lung.

R. Doll and A. B. Hill, "Smoking and Carcinoma of the Lung," *British Medical Journal*, 2, no. 4682 (1950), 739–748.

coincidentally, exactly when tobacco consumption was on a steady rise). No similar image appears in the Wynder and Graham paper. Interestingly, it took at least a decade more in the United States than in the United Kingdom for tobacco to be declared a possible carcinogen. This delay was likely due to many factors, but the striking character of the upward slope in cancer deaths and the clear association with smoking habits cannot be denied.

To ask Royall's second question, what should we believe? There are three issues to consider here: study design, potential confounders, and errors of sampling. The above-described studies are "retrospective"—they involve identifying "outcome" cases, matching them with "controls," and determining, via interviews, what exposures might have led to these outcomes. Such retrospective studies are subject to a variety of known confounders: recall bias (patients may be more likely to exaggerate exposure, given outcome), interviewers' bias (scaling up the smoking habits in lung cancer patients), selection bias (the potential problems associated with identifying appropriate cases and controls—e.g., choosing patients in hospital as controls may result in a control group with lower rates of smoking than the average in the population as a whole), as well as confounding (other possible causal explanations or "common causes"—e.g., atmospheric pollution having to do with locale or occupational exposure to dust or other carcinogenic materials). Both sets of authors discussed a variety of these sources of bias and possible confounders and argued that they could not explain the strong association. For example, Doll and Hill considered whether better methods of diagnosis might have resulted in the dramatic rise in deaths attributed to lung cancer, but argued that the similarity of rates of increase in places with and without the best diagnostic facilities argued against this, citing epidemiological work by Kennaway and Kennaway (1947) and Stocks (1947). However, Doll and Hill did not need to imagine possible confounders.

Confounding causes were suggested by both private interests and other physicians and researchers. Notably, R. A. Fisher (1957) published a series of letters in the same journal (*British Medical Journal*) in which Doll and Hill's paper was published. The letters were typical of Fisher's acerbic style, accusing the authors of "political rhetoric" and "creation of frantic alarm." He also published two letters in *Nature*, in 1958, offering something more than acerbic wit. He argued (infamously), that both smoking and lung cancer might be linked by a common cause: namely, a genetic predisposition to both. He even went so far as to suggest a method by which such a common cause could be discovered—twin studies—and identified at least one study suggesting such a link. Citing a study in Germany by von Versucher on monozygotic versus dizygotic twins, Fisher argued that shared hereditary appears to play a significant role in one's predisposition to smoke (monozygotic twins show greater propensity for sharing smoking habits than dizygotic twins) and, moreover, claimed (or supposed) that "genotypically different groups would be expected to differ in cancer incidence." He also claimed that these twin studies would shed light on the oddity that pipe and cigar smokers were more likely to develop cancer than cigarette smokers, i.e., those who inhaled. Fisher made much of this particular observation, in a second letter to *Nature*, in August 1958. To quote Doll and Hill's (1950) study:

> All patients who smoked were asked whether or not they inhaled, and the answers given by the lung-carcinoma and non-cancer control patients were as follows: of the 688 lung-carcinoma patients who smoked (men and women) 61.6% said they inhaled and 38.4% said they did not; the corresponding figures for the 650 patients with other diseases were 67.2% inhalers and 32.8% noninhalers. It would appear that lung-carcinoma patients inhale slightly less often than other patients (χ^2 = 4.58; n = 1; 0.02<P<0.05). (744)

What is to be made of the fact that inhalers were slightly more common among patients with other diseases than cancer patients? As Doll and Hill point out, "The difference is not large, and if the lung-carcinoma patients are compared with all the other patients interviewed, and the necessary allowance is made for sex and age, the difference becomes insignificant (χ^2 = .19; n = 1; 0.50<P<0.70)" (744).

Moreover, this observation seems to be a red herring, given the significantly stronger association with smoking and lung cancer in general. However, Fisher takes this oddity as suggesting that "an error has been made, of an old kind, in arguing from correlation to causation." He makes much of the possibility that genetic predisposition toward different smoking habits and associated variability in disease predisposition could explain the association, and claims that inhaling might have a "prophylactic" effect:

> There is nothing to stop those who greatly desire it from believing that lung cancer is caused by smoking cigarettes. They should also believe that inhaling cigarette smoke is a protection. To believe either is, however, to run the risk of failing to recognize, and therefore failing to prevent, other more genuine causes. (Fisher, 1958, 596)

In Fisher's typical dry style, he is pointing out an important fact: correlation is not causation. This objection was echoed again and again in criticisms of the epidemiological work on cancer and smoking, particularly by industry representatives. However, as evidence accumulated, it became more and more difficult to adopt this "rationally skeptical" stance. In 1954, Doll and Hill published a second, prospective cohort study, based on interviews with English physicians. The advantage of prospective over retrospective studies is that the latter are not as subject to recall bias as the former. In 1951, Doll and Hill sent questionnaires to 59,600 members of the medical profession in the UK, asking them to describe their smoking habits and to give their name, address and age. Only 41,024 replied; 40,564 of the replies were sufficiently complete to be analyzed, and these respondents were classified by age, sex, and smoking habits. After a lapse of twenty-nine months, the authors accessed the registry of deaths among medical professionals (a public record). Of 789 deaths, 35 were from lung cancer. All those who died were smokers, and there was a direct correlation between amount smoked and death from lung cancer. Deaths from other cancers and respiratory disease, as well as cardiovascular disease, showed a similar correlation. Light smokers or non-smokers made up the smallest proportion of those who died.

The advantage of these latter prospective studies is that they suffer from fewer problems of recall or selection bias. Arguably, however, without identification of a "mechanism," one might always subject such research to "rational skepticism." However, as early as the 1910s and 20s, in Japan, work on mice and rabbits had shown that regular application of tobacco tar to the skin could lead to cancerous lesions and death. The first study published in English to demonstrate such an effect was that of Chikamatsu in 1931. However, the work of Japanese researchers' work done in the 1910s, 20s, and 30s, was not well reported (at least not in the English speaking world), so that only in 1957, with the careful work of Graham et al., were the results replicated. Graham et al.'s striking results and the images of dramatic cancerous tumors in the rather pitiful rabbits would turn stomachs (Figure 4.4). However, this research apparently was not enough to convince everyone. One could always object that one does not apply tobacco tar to one's ear and, moreover, that animal studies provide only indirect support for similar mechanisms in humans. Such arguments became less and less common as evidence accumulated.

This takes us to Royall's third question: What should we do? In 1959, the US surgeon general (then LeRoy Burney) published a statement in the *Journal of the American Medical Association* stating that the Public Health Service believed increased rates of lung cancer death to be due to smoking and that smoking was associated with an increased chance of developing lung cancer (1959). In 1964, the advisory committee of the surgeon general (then Luther Terry) published a comprehensive report arguing that there was a clear link between cigarette smoking and lung cancer. Terry appointed an advisory committee of ten prominent scientists—specialists in lung physiology, chemistry, cancer genetics, pathology, pharmacology, and surgery—who met nine times over nine months at the NIH's National Library of Medicine. They reviewed over 6,000 articles in 1,200 journals and drew upon evidence and testimony from 155 biologists, chemists, epidemiologists, biostatisticians, and mathematicians (Surgeon General's Report, 1964). The evidence was based on epidemiological work, as

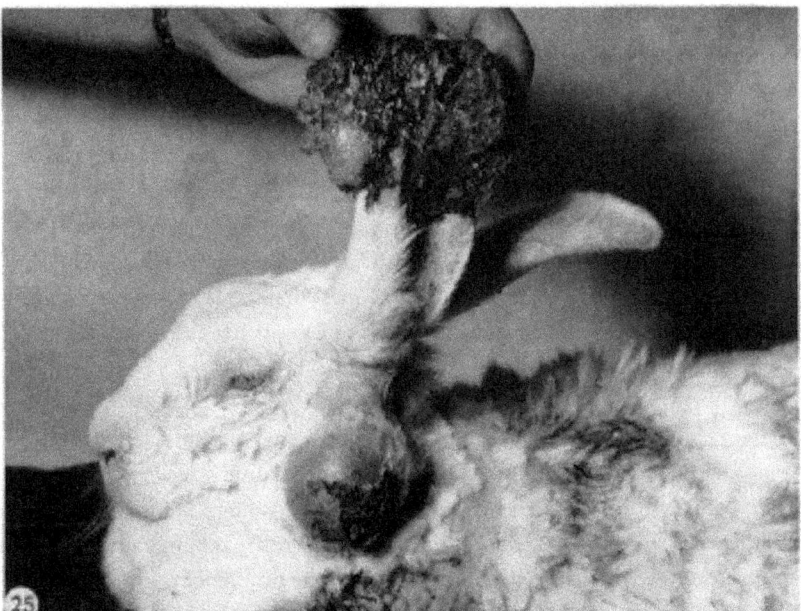

FIGURE 4.4 Skin cancer produced in rabbit ear with tobacco smoke condensate. Original studies conducted by Katsusaburo Yamagiwa and Koichi Ichikawa, "Experimental Study of the Pathogenesis of Carcinoma," *Journal of Cancer Research*, 3, no. 1 (1918), 1–29.

well as animal studies, clinical studies, and autopsy studies. At this point decades of work had been undertaken by hundreds of investigators on tobacco use and its links to cancer; the report was unequivocal on the link between the two and caused a firestorm of publicity, as well as (rather ineffectual) attempts at regulation of tobacco advertising on the part of Congress. It was only after this document was published that the consumption of cigarettes began to decline in the United States. Despite this, the tobacco industry continued to challenge the link, making claims of "safer" filters and low-tar cigarettes well into the 1970s, claims that were not successfully challenged in court until 1987. Even then, Rose Cipollone was found by the jury to be 80% responsible for her smoking habit and was awarded only $400,000 in damages, barely enough to cover legal expenses for her widower (she died of cancer during the trial) (see, e.g., Mukherjee, 2010).

The case of establishing the link between tobacco and cancer is interesting because it parallels the development of professionalization in epidemiology; scientific societies, journals, methods, and general standards of evidence were established over the course of 1950–1970. In 1965, Sir Bradford Hill published a paper that became a locus classicus of epidemiology: "The Environment and Disease: Association or Causation?" Hill outlines a variety of "aspects" or "characteristics" of an association between some environmental exposure and disease relevant to determining whether we can say that the association is "causal"; he concludes with the following striking comments:

> What I do not believe . . . is that we can usefully lay down some hard-and-fast rules of evidence that must be obeyed before we accept cause and effect. None of my nine viewpoints can bring indisputable evidence for or against the cause-and-effect hypothesis and none can be required as a sine qua non. What they can do, with greater or less strength, is to help us make up our minds on the fundamental question—is there any other way of explaining the set of facts before us, is there any other answer equally or more likely than cause and effect? (1965, 299)

Hill is careful to say that his "aspects" are not "necessary and sufficient" criteria—they are, at best, rules of thumb: a heuristic for determining whether some association is likely to be causal. Hill is overtly critical of the use of the chi-squared test and other tests of statistical significance, as "there are innumerable situations in which they are totally unnecessary—because the difference is grotesquely obvious, because it is negligible, or because, whether it be formally significant or not, it is too small to be of any practical importance" (299). One can read the paper as a reply to Fisher about the decisiveness of statistical tests in determining causation: "the magic formulae" of tests of significance, Hill argues, can "serve as guides to caution" but *should not be taken as a substitute for other considerations.* Additional factors worth considering when assessing whether an association is in fact causal are, in Hill's view:

- Strength
- Specificity
- Consistency
- Temporality
- Biological gradient
- Plausibility
- Coherence
- Experiment
- Analogy

This paper by Hill has been one of the most frequently cited, disputed, and discussed in the epidemiological literature, so it is worth pausing to analyze exactly what he meant by these factors and how they were meant to be applied. They are particularly useful when one considers the case of tobacco, which Hill does at several points in the 1965 piece. For instance, *strength* of association, he argues, is a strong indication of causality. He discusses the case of scrotal cancer among chimney sweeps, as well as the case of smoking and cancer, and argues that when the association is as strong as it was in these cases, it seems much more likely that it is causal than that it is due to common cause. Of course, he points out that strength of association by itself cannot rule out common causes, but "such a feature of life" would have to be "so intimately linked with cigarette smoking and with the amount of smoking that . . . [it] should be easily detectable" (296). In other words, if there is no other imaginable cause that could explain the link, a strong link should weight our judgment in favor of a causal association.

But what is strength of association? Hill discusses two measures of strength of association: what are now called "relative risk" and "absolute risk." The relative risk is a ratio: the incidence of disease among exposed over unexposed populations. For instance, chimney sweeps were as much as two hundred times more likely than the unexposed population to acquire scrotal cancers, and smokers were from ten to thirty times as likely to die from lung cancer than non-smokers, depending on duration and extent of smoking habits. "Absolute risk" is a more modest number—it is the absolute difference between death rates from lung cancer in smokers versus non-smokers, which range from 0.57 to 2.27 per 1,000 for smokers, depending upon how many cigarettes they smoke daily, versus 0.07 per 1,000 for non-smokers. Hill is careful to point out that strength of association is neither necessary nor sufficient for causation; weak association can still be causal (as in the case of low probability of developing meningococcal meningitis after infection from menigococcus), and strong association can be due to common causes. Philosophers will be familiar with the case of the barometer and the storm; the common cause in this case is atmospheric pressure. So Hill is well aware that strength of association is neither a necessary nor sufficient condition for causal association; it is simply one consideration among many.

Consistency is a measure of whether an association has been "repeatedly observed by different persons, in different places, circumstances and times" (296). Here Hill is referring to a feature of evidence that one might refer to as "robustness" or "stability": where the same association is found in so many places and times, across so many different background conditions, one might have more reason to suspect that it is causal. This is because if an effect is shown in very different contexts, confounders unique to one population or one context are reasonably regarded as unlikely. Again, this factor was in place in the case of smoking and lung cancer; in different continents (US and UK), locales (major cities versus rural populations), and populations (physicians as well as non-physicians), a very strong association was discovered. However, Hill is also circumspect about the epistemic warrant for this factor: "No number of exact repetitions would remove or necessarily reveal that fallacy" (in this case, the assumption that two populations are identical apart from suspected causal factors). And he also points out that "when repetition is absent or impossible," (297) there are some cases where a single study can provide sufficient evidence for causation. The example he discusses is intriguing; there was a significantly higher rate of deaths from nasal and lung cancer among nickel refinery workers before protective measures were adopted in the factory in 1923. No excess cancer deaths in this population were found after this time. Such a remarkable shift might be taken to count as sufficient evidence, Hill argues, for a causal association, insofar as this case study resembles a quasi-experimental study—the intervention appeared to result in a marked change in the incidence of this particular (and rare) cancer.

Specificity is a measure of whether some association is limited to a specific kind of exposure and a particular kind or site of disease. Here the example of scrotal cancer and chimney sweeps comes to mind; such a rare cancer among a very unusually exposed population is an example of a case where the disease–exposure link was quite specific. However, again, Hill is circumspect; he remarks that some exposures are related to a variety of diseases (poorly processed milk can lead to many kinds of infection), and most diseases are likely to be multicausal in origin. However, Hill argues that "if specificity exists we may be able to draw conclusions without hesitation" (1965, 297).

Temporality means that the cause must precede the effect. Hill gives specific examples of cases where disease has a slow onset, and it may not be possible to determine whether an occupational or environmental exposure was a cause or simply strongly associated with another cause.

Association should reveal a *biological gradient*, or "dose–response curve." In other words, if more exposure is linked to higher rates of disease, one might have more reason to suspect that the two are causally linked, though Hill also remarks that other, more complex dose–response curves for various exposures are possible.

Biological *plausibility* is an important feature of the causal relationship. "But," Hill points out, "this is a feature I am convinced we cannot demand," for "what is biologically plausible depends upon the biological knowledge of the day." In other words, given that our biological understanding is always incomplete, we should not rule out a biological link if we cannot yet imagine one, given our current biological understanding.

Coherence refers to the fact that if the causal interpretation of our data "seriously conflicts with the generally known facts of the natural history and biology of the disease," (298) then we ought to reconsider the suspected causal association. The known histopathological evidence of effects on the bronchial epithelium of smokers leant greater, not lesser, strength to claims of causal association. However, Hill remarks that the fact that Snow's epidemiological observations of the association of cholera with the Broad Street pump should not have been dismissed simply because the mechanism of infection was then unknown.

Experiment (or what he calls "semi-experiment" (298)), can occasionally be appealed to. The case of the nickel workers was a kind of natural experiment; after the protective measures were taken, cancer rates decreased. The case of smoking is another; over the past several decades, smoking rates have declined, followed by a marked decline in cancer of the lung, particularly in men. This is a kind of "experimental" evidence in that one can measure the effects of an intervention, though in this case, only after the fact.

Finally, Hill remarks that one can judge by *analogy*. For example, when two infectious agents or drugs are very similar, either in mechanism or in chemical structure, we may have good reason to think that their behavior might be similar in the body.

On Hill's view, it seems fairly clear that tobacco is a significant cause of lung cancer; the association between tobacco use and increase in lung cancer risk is strong; it has been found in different populations in different environments and presumably with respect to different suites of risk factors. Moreover, as of 2000, experimental or quasi-experimental evidence shows that as rates of smoking have gone down, so too has death from lung cancer. Finally, the animal studies on tobacco tar and tumor induction, as well as the evidence from dissections of smokers' lungs, suggested that there is a plausible biological mechanism—the tar in cigarettes causes inflammation, scarring, and ultimately changes in the tissue of the lung that lead to cancer. Until relatively recently, exactly how this occurs was unknown (Hecht et al., 1988; Hecht, 2002). Tobacco smoke contains a wide array of compounds that are deleterious to health; some of these compounds, such as 4-(methylnitrosamino)-1-(3-pyridyl)-1-butanone (NNK) and N'-nitrosonornicotine (NNN) are nicotine derivatives and are highly carcinogenic. These molecules can form adducts with cellular DNA, leading to mutations in genes like *Ras, p53*, and *Rb*, each of which control cell birth and death.

Mutations in these genes are common in many cancers—from lung to breast and colon cancers (Weinberg, 2007). These genes and their role in cancer were not discovered until Bishop and Varmus's Nobel Prize–winning work in the 1980s (discussed in Chapter 6), and so establishing the exact mechanism by which compounds in tobacco cause cancer was not possible until the early 1980s. However, long before this, warning the public of the risk of tobacco was (arguably) warranted. This suggests that something less than a full or detailed mechanistic causal explanation can provide sufficient grounds for action.

In 1965, in Hill's view, the evidence was more than sufficient to warrant public policies warning smokers of risks to their health. Hill argues that we can and should take action, even if we are not absolutely certain of a causal link. His list of factors to take into consideration, moreover, are not intended to be necessary or jointly sufficient conditions, but only matters worthy of reflection. Hill knew well from personal experience that the best one can do in epidemiology is provide a strong case for a causal association by describing strong associations, ruling out confounders, and having at least some plausible mechanistic explanation in mind. However, despite his careful qualifications about the variety and nature of evidence, epidemiologists for some decades have been critical of Hill. Historical context may provide some clue as to why.

The professionalization of epidemiologists coincided with the rise of the "molecular revolution" in biology. Only three years after Wynder and Graham's 1950 study and two years after Doll and Hill's work on tobacco and smoking, Watson and Crick published their 1953 paper that won them a Nobel Prize along with Maurice Wilkins in 1963 "for their discoveries concerning the molecular structure of nucleic acids and its significance for information transfer in living material." With the subsequent rise of molecular biology in the 1970s and 1980s and the human genome sequence completed in 2001, arguably, the standards of evidence for establishing causal links in cancer were shifted to the molecular level. In other words, just shortly after epidemiologists became professionalized, launching journals and developing graduate programs in public health and epidemiology, advances in molecular biology and genetics began to take center stage in the scientific community, attracting far more funds and prestige. Molecular biology and genetics are often considered closer to the gold standard or experimental (interventionist) ideal, results are often more precise and quantitative, and observation is considered more "direct" (though what one means by "direct" is far from clear). Thus, a good deal of epidemiologists' framing of their own research, and debates about warranted evidence, are set against the background of fields such as genetics and molecular biology and experimental "mechanistic" ideals for establishing causation.

In part as a result of this transformation in biology, molecular epidemiology arose in the 1980s as a way to fill in the "black box" of traditional epidemiology. Molecular epidemiologists can confirm claims of exposure through the identification of biomarkers. Further, they can even establish the dosage and mechanism of action of a purported carcinogen. Moreover, they have extended genetic medicine into epidemiology by identifying genetic variations that contribute to the risk of disease or the likelihood that a disease will respond to various treatments, either because of inherited variation in drug metabolism or genetic features of a particular tumor. Biomarkers can help identify which carcinogens are acting in which populations and make the link between exposure and disease more concrete.

With this shift in prestige and funding to genetics and molecular biology, there has been a shift in the rhetoric of epidemiologists about evidence for causation. Many see themselves as attempting to match or closely approximate criteria of evidence for causation of cell and molecular biology, and use talk of counterfactuals and intervention to reframe new statistical tools or models used in epidemiological research. However, in my view, this is a mistake; the appropriate model for evidential reasoning in epidemiology is simply not the same as the model appropriate for cell and molecular biology or, for that matter, clinical medicine. Reiss (2015) and Broadbent (2011c) make similar arguments; and, indeed, some epidemiologists have also resisted this assimilation of evidential reasoning in epidemiology to evidential reasoning in clinical medicine. Broadbent (2011c) has identified what he calls two extremes of a continuum of views in epidemiology: "the mechanistic" stance and the "black box" stance. The former view holds that establishing a causal association requires the identification of a "mechanism." Just as Illari (2011) notes, however, there is some vagueness about this standard that is worth remarking upon. How detailed need one's understanding of "the mechanism" be to warrant claims about causal relations? Must one know exactly which genes are mutated by which compounds and how? Broadbent contrasts the mechanistic with the "black box stance," which holds that classical analytic epidemiology (of the sort conducted by, e.g., Doll and Hill or Wynder and Graham) is sufficient to establish causation. It appears that the black box stance is losing ground in the current push toward molecular methods and the identifications of biomarkers for both disease risk and/or exposure. Classical methods and tools have been overshadowed by shifts toward methods that deploy the "harder" laboratory sciences, such as molecular biology, bacteriology, chemistry, and virology. Morabia (2011), an epidemiologist trained in the tradition of Doll and Hill, bemoans this competition for authority and resources in his paper "Until the Lab Takes It Away from Epidemiology." Establishing probabilistic associations at the population level is indeed a challenge. Consider the problem of common causes. Suppose smoking raises the chance of myocardial infarction; suppose also that smokers in some population are more likely to exercise, which in turn lowers the chance of myocardial infarction. So smoking appears to decrease the chance of myocardial infarction. Epidemiologists cannot ever expect to have perfectly homogeneous populations. If a necessary condition of establishing causality is that we somehow have knowledge of all possible confounders, establishing causation in epidemiology would be simply impossible. Epidemiologists are well aware of this problem; indeed, methods of correcting for confounding are in some ways attempts to "homogenize" the class of individuals subject to some exposure. However, epidemiologists can never be absolutely certain that a Simpson's paradox problem is not in play. In contrast to other biomedical scientists, classical epidemiologists rarely, if ever, establish connectivity or temporal continuity between purported causes and effects, and almost never "universality," or the non-exceptionless linking of cause and effect (Schaffner, 1997).

Molecular or genetic epidemiology addresses a rather different question than classical epidemiology. Molecular biologists investigate mechanisms linking exposure and disease, whereas classical epidemiologists try to find out if any such link is plausible or likely and, more important, if it is worth acting on in service of public health. The approaches are not at direct odds, but they address different questions: the first asks how or why, and the second, whether there is a link between exposure and disease. Such information can be had without a detailed understanding of mechanisms, provided one has ruled out *most likely or plausible* confounding causes, among other considerations mentioned above. For those who argued that a mechanism is a necessary condition for "confirmation" or "explanation" of the causal link between some exposure or disease, it would have been necessary to wait until the early 1990s for the link between smoking and lung cancer to be established. However, with Hill, we may wish to act before this kind of mechanistic understanding is available. Even if detailed knowledge about which individuals are more genetically susceptible to cancer might allow for more fine-tuned interventions, it seems that we by and large might be better safe than sorry. As Hill remarked at the close of his 1965 paper, "Who knows, asked Robert Browning, but that the world may end tonight? True, but on available evidence, most of us make ready to commute at 8.30 next day."(12) Let us turn to a more contested case, however, than smoking and lung cancer.

4.4. Ionizing Radiation: The Downwinders

[I]f you are going to become a victim of military testing, your chances of getting justice are best if there are lots of other victims you can contact and organize. If you are alone or in small numbers, or if a lot of time has passed and your fellow victims have moved away, the responsible powers will roll right over you. (Ward, 1999, 102)

Between 1951 and 1962, the United States conducted 928 atmospheric and underground nuclear tests in the Nevada Test Site (DOE/NV-209-REV 15, 2000). One hundred of these tests were atmospheric (aboveground) and 828 were underground, as aboveground testing was banned after 1958. Fallout from 26 of these tests conducted between 1951 and 1958 was carried by winds into southwestern Utah from the Nevada test site, as documented by the US Defense Nuclear Energy and Energy Research and Development Administrations' maps of the fallout (Dunning, 1959). Rural families living in southeastern Nevada and southwestern Utah could see the clouds and, in some cases, felt the blast or felt ash raining down when the wind sent fallout east. In some cases, families in southeastern Nevada and southwestern Utah made road trips or deliberately got up early to sit outside and watch the blasts; schoolchildren in some cases were permitted to go outside and play immediately after the detonations.

As Ward points out in the epigraph to this section, establishing causation, particularly for small populations in remote areas, where the impact is felt only decades later, is enormously difficult. In fact, one can quantify exactly how difficult. In *Environmental Health*, Moeller ([2005] 2009) opens his chapter on epidemiology with a table describing the "size of exposed population group and radiation dose required to detect an increase in total cancer mortality, assuming lifetime follow up" (51). The table is rather bleak, or would certainly appear so, to any epidemiologist attempting to establish greater cancer risk due to fallout from the aboveground tests from 1951 to 1958. Essentially, in order to "prove" a significant increase in cancer mortality (or, more accurately, provide statistical evidence for such an increase, where $p = .025$), one needs an enormously large sample. According to Moeller's table, at a mean dose of 2.5 mSv (millisieverts), one would need to demonstrate 1.9 excess cancers per 10,000, which would require a sample size of 32,000,000. To give some perspective on this, the average background radiation over one's lifetime is about 2.5 mSv. The more acute the dosage, in other words, and/or the larger the sample and number of "excess cancers," the more likely it is that one can establish that radiation was responsible. Mayo (1988) makes a very similar point with respect to the case of formaldehyde; the ability to detect carcinogenicity in the study conducted and cited by the DuPont company was so weak that reliance upon a study by the Environmental Protection Agency (EPA) practically guaranteed that formaldehyde would not be found carcinogenic. Exactly because the effects of many carcinogens are so weak and can take decades to manifest themselves, it is all too easy to reject the hypothesis that an entity is carcinogenic. Mayo argues on these grounds that an error-statistical approach can make transparent to regulators exactly how weak our evidence is, and thus we are in a better position to assess the relative value of evidence in making practical decisions.

The case of the Downwinders is, arguably, a very weak one. But this is in part because the sample size necessary to establish anything like a very strong association would be—effectively—impossible to get. Thus, the environmental epidemiologists needed to build a case from a variety of evidence and a relatively small sample—it was, in some sense, a numbers game. One needs, first, evidence of an excess of cancers in the population exposed to radiation fallout. Second, one needs to provide reasonable grounds for extrapolation back from high-dose (acute) studies if numbers are less than satisfactory—in this case, based on taking samples of radiation-affected structures in the environment and extrapolating backward to likely exposures in children. When and whether such extrapolations are warranted depends upon some assumptions about the linearity of dose–response curves.

Let's turn to Royall's first question: What was the evidence? It will be useful to describe the details of the epidemiological study and the "numbers game" at play in this case. The "gray" (Gy) is a measure of "absorbed dose of radiation" (D). The sievert (Sv) is called the "equivalent dose of radiation" (H). Both the gray and sievert have dimensions of energy per unit mass, 1 joule/kg; 1 "rad" is 10 milligray (mGy). The equivalent dosage to a tissue is a multiple of the absorbed dose, measured in Gy, by a weighting factor (W_R). The relation between absorbed dose D and equivalent dose H is thus

$$H = W_R \bullet D$$

The excess dosage experienced by the Downwinders was in the range 0.5–30 Gy, depending on age and diet. For instance, children experienced higher dosages, if they drank milk from cattle or goats that had been exposed to the fallout, and so had concentrated irradiated fallout in their milk. To give sufficient evidence by the lights of Neyman-Pearson standards of error that a given cancer is due to radiation exposure, sample sizes would need to be on the order of 107 to definitely establish excess cancers, according to an image from an article by Brenner et al. (2003) (see Figure 4.5).

What, then, do we have of evidence in this case? There are three main lines of evidence. First, information about the tests themselves, wind speeds, and levels of radiation is now a matter of public record. Second, epidemiological evidence concerning the cohort affected lent support. And third, there was indirect evidence from impacts on animals grazing in the area, as well as from sampling wood in buildings at the site of exposure.

Two epidemiological studies in the 1970s and early 1980s provided the initiative, first, for a legal case brought in 1984 on behalf of the Downwinders and, second, for a bill sponsored by Senator Orrin Hatch of Utah that provided compensation to the same. I interviewed Joseph Lyon, an epidemiologist at the University of Utah, who published some of the first studies on childhood leukemia among survivors, in the *New England Journal of Medicine*, on February, 22, 1979. The following is a portion of the transcript of our interview.

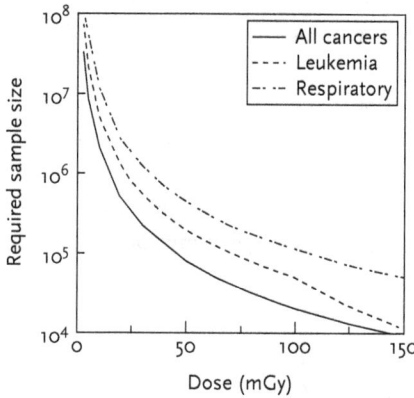

FIGURE 4.5 Sample sizes needed to establish association between exposure and disease, in this case leukemia and radiation exposure.

From D. J. Brenner et al., *Proceedings of the National Academy of Sciences*, 100 (2003), 13761–13766.

Our paper was published in Feb. 22, 1979.

The feds finally funded the fallout study to confirm the leukemia findings. We went back and found much more thorough information about who got leukemia and where they lived. We used LDS [Church of the Latter Day Saints] records, so we could track where they lived through their whole lives. Then we tried to get a clear sense of how much radiation exposure they actually had. We finally went through a whole host of ways—it forced the federal government to locate, review, and summarize old records of fallout radiation from across the entire U.S., something that had not been done before.

Before this paper, there were no electronic files on how much radiation there was. There was no uniform database.

... The heaviest exposure was May 19, 1953. It was Shot Harry ... [one of the last aboveground tests, conducted May 19, 1953], because of unsafe wind speed, the AEC [Atomic Energy Commission] at the Nevada Test Site did not want to explode the bomb at the scheduled time (about 5AM in the morning), because of unstable weather—but Eisenhower and the White House insisted the test go ahead as scheduled.

... We found that 60–85% of exposure came from that one shot: Shot Harry ... over the next seven to eight days from the milk that people were drinking ... it was private backyard cows.

Cows absorb it, concentrate it, and put it in their milk; iodine loves milk, it's highly soluble. The only other source was a vegetable garden.

Milk has always been the big bugaboo for iodine exposure. We finally were able to get reasonable estimates when the feds finally funded in the 1980s. The kids who died from leukemia between 1955 [and] 1960 were probably out playing in the sand boxes and the schoolyard.

When we finally got that information, it started to make more sense—it was a very acute exposure. Some of the radiation people on campus already knew it, they mentioned Shot Harry, and the concerns about the large amount of exposure from that single shot.

Physicists are used to dealing with very precise measurements—they are used to many decimals. We epidemiologists are saying that people were 2–3 times more likely—there's been a lot of work on misclassification, showing that if people are randomly sampled, then you underrepresent the effect. Some will say that the Dr's don't know how to diagnose leukemia. Until 1965, there was no effective treatment. Mortality was 100% and 3 months of life post diagnosis. Here we had at U. of [Utah] Max Wintrobe, world famous hematologist. They knew how to diagnose leukemia. Then we got into the more absurd ones [purported confounding causes]: background radiation. So, we counted houses made out of brick versus wood. There were a lot more made out of wood in St. George. The feds were sending out a group of people every three months—they were coming up with explanation after explanation, and we tried to counter it. We had been through background radiation, diagnoses, every possible bias that they could come up with . . . . George Land was a statistician at this meeting. We had been assaulted after two hours—every other explanation. Land—in this library in the U., said: "it seems to me that the simplest explanation for all the data that we've looked [at] is the fallout— it covers every fact that we've seen."

All the other feds were silent, mumbling—suddenly shut down. He carried a lot of weight because of his intellectual ability and honesty. At that point I thought, Yeah we've won. And isn't it interesting how science gets done? That was the last serious challenge.

As this statement illustrates, the evidence was contested, and debates were politicized. However, the gradual accumulation of several independent lines of evidence cinched the case for a causal relation between the excess incidence of leukemia and thyroid cancer and ionizing radiation.[1] Lyon's studies were a case-control and cohort study, showing elevated risk of leukemia and thyroid disease in children exposed to radiation through either external or internal exposure from the Nevada tests. The first, a case-control study, examined rates of leukemia deaths in children in the region. The study showed that leukemia was approximately 2.5 times more likely in populations exposed to the radiation fallout than in a control group. However, the sample was small (see Table 4.1).

---

1. One source of evidence is simply based on a general understanding of the health effects of radiation and how they vary by wavelength or frequency. The electromagnetic spectrum ranges from ionizing to ultraviolet, to visible and infrared light, to microwaves and radio waves. Ionizing radiation is radiation caused by X-rays, emissions from nuclear reactor accidents, and nuclear weapons detonations; this is a short-wavelength, high-frequency radiation that has the ability to remove electrons from atoms or molecules. The positively charged molecules are highly unstable and can produce chemical changes in the surrounding environment—such as free radicals. Such radiation can penetrate tissue easily, causing damage to cells ranging from critical irreversible chemical changes to DNA to "bystander" effects—effects in cells in close proximity to the cells directly affected. At high doses, ionizing radiation causes death; at lesser doses, it inhibits mitosis; and at any dose, it can cause alterations in the genetic material of the cells. This leads to a range of effects over the longer term, but the cells most sensitive to radiation are those that divide frequently—bone marrow cells that give rise to blood cells and platelets, and the cells that line the small intestine and stomach. Survivors of the Hiroshima and Nagasaki bombings, for instance, experienced excess mortality from leukemia and other cancers, including breast, lung, stomach, and thyroid cancer, over the twenty-five years following exposure (Brill et al., 1962; Isimaru et al., 1966; Moriama, 1978). Also, radioactive iodine persists in the food chain (e.g., fallout deposited on grass or soil, when consumed by cows or goats, accumulates in milk) and when ingested, particularly by children, can lead to diseases of the thyroid. There were, for instance, large increases in thyroid cancer in children living close to Chernobyl within a few years of exposure.

Table 4.1 Standardized Leukemia Mortality Ratios (SMR) for the High-Exposure Cohort as Compared to the Low-Ex-posure Cohort for Utah and High-Fallout and Low-Fallout Counties.

Area	Sex	Observed Deaths*	Expected Deaths†	SMR‡	Confidence Interval§ low	upper
Utah	M	89	57.8	1.54¶	1.00	2.37
	F	95	74.1	1.28	0.93	1.76
Totals		184	131.9	1.40‖	1.08	1.82
High-fallout	M	16	5.6	2.88	0.96	8.60
counties	F	16	7.5	2.12	0.63	7.11
Totals		32	13.1	2.44¶	1.18	5.03
Low-fallout	M	73	52.2	1.40	0.84	2.35
counties	F	79	66.7	1.19	0.87	1.63
Totals		152	118.9	1.28	0.97	1.69

*Observed deaths were those occurring in the high-exposure cohort.

†Expected deaths were generated by application of age-specific mortality rates to person-yr at risk in the high-exposure cohort.

‡Tested by the Mantel-Haenszel procedure controlled for age & sex.

§Approximate confidence intervals after Miettinen.I'

¶P<0.05.

‖p<0.01.

Note: Sample sizes (and exposures) are small, so the extent of variation across the groups is difficult to detect.

Source: Lyon et al., "Childhood Leukemias Associated with Fallout from Nuclear Testing, *New England Journal of Medicine*, 300. no. 8 (1979), 397–402.

As is apparent from Table 4.1, in the state of Utah there were 184 deaths from leukemia among children of ages 0–14 between 1959 and 1967. Among those in the "high-fallout" counties, mortality was two to three times what might be expected in a population of similar size. In other words, the excess deaths were concentrated among those born in the high-fallout counties between 1951 and 1958. There was also a decline in deaths from leukemia among children born after the high-fallout period. However, the absolute number of deaths in these populations was small. "Only" thirty-two died of leukemia in the high-fallout counties. As Lyon et al. report:

Although this difference was not statistically significant, it was in the same direction and, combined with the high-fallout counties, produced a significant excess in the state rates for the high-exposure cohort as compared to the low-exposure group. For the other childhood cancers in the two cohorts, no consistent pattern in relation to fallout exposure emerged. No significant excess for these cancers was noted in the high-exposure cohort as compared to the low exposure group. (1979, 398)

In other words, there was a difference between the high- and low-fallout counties in leukemia deaths, particularly among children born in 1951. Yet this difference was not statistically significant. What are we to make of this?

Let's consider Royall's third question: What should we do? Recall Hill's critique of Fisher: the "magical formulae" of statistical significance are not, by themselves, a necessary and sufficient guide to causal associations. This is in part why Lyon and his group also collected samples of wood from buildings in the relevant counties to detect radiation levels, and thus likely exposures, of children. Perhaps more relevant to this case, however, one should *expect* that it would be enormously difficult to identify a statistically significant difference among high- and low-fallout counties, insofar as the populations were small and children in both areas were exposed (though to different extents). As later work on nuclear fallout from the testing has shown, excess cancers could be caused by fallout from the Nevada Test Site spread by wind and weather as far as Idaho and Chicago. Once we scale up to the United States as a whole (and with better data about dosage from fallout as well as the distribution over the United States, based on maps from National Cancer Institute), it is a bit easier to measure statistically significant causal associations. Again, though, this is a numbers game; one report estimates excess thyroid cancers, particularly among children, ranging from 230,000 to 440,000 (Hoffman et al., 2002). The large range of numbers is due to the extent of uncertainty about levels of exposure, due to lack of knowledge about how much commercial milk children in various populations consumed, whether the milk was from a commercial retailer or a backyard cow, and, of course, uncertainty about the spread and extent of fallout. However, according to Hoffman et al., "If the present criteria for compensating exposed and sick radiation workers and atomic veterans were to be applied to the public, then people throughout the U.S. who have thyroid cancer or a non-cancer neoplasm would be eligible for medical care and compensation if they were children when exposed to Cold War era 131I and if their diets were composed of fresh milk from either cows or goats" (2002, 746).

In other words, we are all Downwinders. Or, more precisely, all US children born in 1951 or shortly thereafter who consumed milk, especially from backyard cows, are Downwinders. The reason for this is that the legally established criterion for establishing eligibility for compensation is that the upper 99th percentile of the estimate of causation must exceed a value of 50%, and this criterion is met in such cases. What this suggests, at minimum, is that Lyon's data may underestimate the risk of cancers among Downwinders, if even the "low-fallout" counties might have been subject to similar risk. Apart from this, however, the case illustrates that our judgments about causation cannot come down to statistically significant outcomes alone; independent evidence from a variety of sources, indicating a plausible mechanism and supporting claims about level and extent of exposure, shaped judgments about action. In part, the decision to compensate the Downwinders was a political decision, but arguably it was also founded on an inference to the best explanation, a judgment that was warranted, given the variety and the weight of several independent lines of evidence.

4.5. Summary: Evidence and Causation in Epidemiology

There are a variety of object lessons to be gleaned from the discussion of environmental epidemiology of cancer in this chapter. First, perhaps rather obviously, epidemiological detective work is challenging: exposure is often difficult to establish in the first place, effects of exposure are diffuse, and, of course, cancer itself can take decades to develop. Thus, there are simply practical limitations to establishing causal links in the environmental epidemiology of cancer. Evidence has to be pieced together from a variety of disciplines that focus on mechanisms of action in humans and other animals. Claims about evidence for causation may involve appeals to less than purportedly ideal or indirect evidence. This is not evidence of the sort imagined by clinical investigators in randomized clinical trials. Epidemiologists cannot manipulate causal variables in the way researchers in a clinical setting can, nor can they control for all possible confounders. These differences do not reflect idiosyncratic or arbitrary standards, but are in part due to the particular kind of phenomena studied. Epidemiologists measure patterns of causal association in populations over time, sometimes spanning decades or generations. This is a very different kind of data than the sort relevant to causal relationships between the administration of drugs and health outcomes like clearing bacterial infection.

Some philosophers argue that in reasoning about cancer's causes, we use inference to the best explanation, reasoning by extrapolation or some combination of the two (Cranor, 2006, 1997; Steel, 2007). There is, of course, a long history of debate about the character of such inferences and whether they are "mysterious" or can be made "systematic" in various ways. While much can be said about the conditions under which such inferences are optimal, this much I will take for granted. There are better and worse such inferences; the ideal inference is one that draws upon the widest array of independent evidence, where the evidence consistently points to a specific and predictable outcome and (ideally) where there is at least some plausible mechanistic description of the process yielding cancer for some proposed cause—in other words, Hill's standards.

What sense of "variety of evidence" do I have in mind? The details of the case will matter, but roughly, in my view, variety of evidence is (more or less) confirmatory when:

(a) Given a causal claim, C, there are a different and independent sources of evidence that raise the probability of that causal relation obtaining.
(b) "Independence" refers to distinct theoretical, experimental, or observational contexts.[2]

In my view, these conditions were met in the cases discussed in this chapter. The central causal core (that both radiation and smoking are associated with cancer) is supported by a variety of empirical evidence from animal studies, as well as long-term cohort and case-control studies. Of course, the evidence underdetermines the conclusion. Does appeal to "mechanistic" bases solve this problem?

No. The Russo-Williamson thesis that knowledge of "mechanisms" is a necessary condition on causal knowledge is ambiguous, at best. What counts as sufficient evidence for a "mechanism" will depend significantly on the particular context in which the information is sought. Moreover, demand for a mechanistic description at the cell or molecular level would seem to discount plausibly relevant evidence of different kinds of causal dependency relationships at other scales and would have delayed important preventive measures that saved countless lives. Smoking was plausibly shown as early as the 1950s to be an important causal factor in lung cancer, even without a detailed mechanistic story about how the lungs are affected. In other words, classical epidemiology is "evidence-based," and we need not await mechanistic accounts to warrant action (for a similar argument, see, e.g., Broadbent, 2011a,b,c). Moreover, toxicologists can use model systems, and often seek analogue mechanisms, for testing the impact of various potential carcinogens (cf., Mayo, 1988; Steel, 2007). Such research provides causal information that is at minimum relevant to setting basic standards of permissible exposure to various chemicals. In combination with other lines of evidence, I would argue, these causal inferences may warrant action.

One worry that this discussion may raise concerns the intrusion of values into scientific decision-making. A fairly commonplace intuition that many people share is that "values must be kept far away from science." In other words, experiment or observation should, where possible, decide empirical questions. We should not believe simply what we happen to *want* to believe or what we happen to value. Scientists ought to strive to be as objective as possible. I agree.

---

2. For a recent discussion of variety and independence of evidence see Schupbach, 2018.

But as the case studies in this chapter suggest, endorsing this ideal does not require eliminating all trace of values. Insofar as we need to establish standards of evidence for rejecting or accepting hypotheses, we need to make choices, and these choices are (in part) value-laden. Does the appeal to values necessarily compromise objectivity? Whether they do, and how much, depend on which values and how they influence our reasoning. Scientific inquiry is not "either" objective or not, but more or less objective, depending upon whether we are transparent about our choices in standards of evidence. Such choices should not be ad hoc, but set out in advance and widely agreed upon given the subject matter and consensus views of scientists specializing in that subject. In other words, our evidential standards and methods will be context-dependent.

Does the context-specificity of evidential standards mean science is less "objective"? No. By way of a simple example, randomized clinical trials are an appropriate method for studying the effectiveness of drugs in treating disease. However, one would not use RCTs to investigate whether smoking causes lung cancer or whether humans and bonobos are more closely related than humans and chimps. This does not mean that epidemiology and evolutionary biology are not objective or don't have standards. It just means that the way evidence is collected and inferences are drawn from that evidence is different for different kinds of scientific questions (cf. Cleland, 2002).

How do we decide whether evidence is decisive, or decisive enough? This is somewhat easier to make precise in the context of frequentist hypothesis testing. According to Mayo (1988), a test is more decisive when it is sufficiently *severe* to rule out the hypothesis we hope to reject or to rule in what we hope to accept, and in many cases, we can quantify the severity of a test. What will count as severe (enough) may depend on the details of the case. It is very difficult to prove that cancer increases risk 1%. It is much easier to disprove that it increases risk 57%. This point can be (somewhat) generalized to the case of epidemiology: a very serious risk factor (tobacco smoke, which increases lung cancer risk significantly above average) is much easier to establish than one that increases risk only slightly (radiation exposure). We need to keep this in mind when assessing the import of the evidence. Requiring high standards of evidence in cases where risks are small or diffuse in space or time is simply setting up an impossible ideal.

Terms like "high" and "small" may seem to be impossibly vague. Are appeals to such terms acceptable when we are considering such questions as what counts as sufficient evidence for causal claims? Mayo argues that there is a sense in which such choices are unavoidable. We can give objective measures of the severity of our tests, but we cannot avoid the value-laden question of where to set the cutoff for severity. In many contexts, we cannot use the kind of objective measures of severity Mayo prefers. Reiss (2015) considers these cases and argues that what determines a good standard of evidence is context-dependent; epidemiology is simply different from hypothesis testing in other contexts. I agree. But I have also argued here that the pragmatics of evidence depends not only on the nature of the data we're gathering, but also on what we wish to use it for. We need to be cognizant of where and how values enter into our methodological choices, and we must also be transparent. That values enter in does not necessarily compromise objectivity. Objectivity is compromised when values are hidden from view and when we make unwarranted claims about strength of evidence.

CANCER FROM AN EVOLUTIONARY PERSPECTIVE

Nothing in biology makes sense except in light of evolution.
(Dobzhansky, 1964)

5.1. Evolutionary Thinking about Disease: A Taxonomy

What does it mean to view cancer in light of evolution? The first thing to note is that not all evolutionary explanations are adaptive explanations; as Fisher said, "Natural selection is not evolution" (1958, vii). Thus, viewing cancer in light of evolution is not exclusively a matter of showing how various aspects of cancer might be a product or byproduct of natural selection. An evolutionary explanation is any explanation that makes appeal to causes of changes in the distribution of variation in a lineage over time—whether genotypic or phenotypic or at the scale of populations, species, families, or genera. Such causes are not limited to natural selection. Drift, or random sampling of variation from one generation to the next, mutation, migration, developmental constraint, horizontal gene transfer, niche construction, and much else besides are causes of change in populations over time. Moreover, selection does not override all other causal factors in evolution; where, how, and how much selection plays a role in a population is contingent on many factors in combination. As Jacob argued (1977), evolution is a "tinkerer"; change over time is constrained by the variation available, how that variation is distributed and expressed over the course of development, and of course how much environments and thus selective conditions change over time. Many contingencies and trade-offs constrain the nature and direction of change in any evolving lineage. This last insight is especially important for determining the explanatory role evolutionary thinking can play in our understanding of cancer.

Contingencies and trade-offs are a central feature of evolution. A trait may be adaptive at one stage in life history but may compromise fitness later in life. Traits that optimize reproductive success may compromise survival. Traits that are optimal for a mother may compromise fitness for her offspring, and the reverse. Traits optimal for the male of a species may compromise fitness for the female, and the reverse as well. All adaptive change is subject to constraints and trade-offs. It is especially helpful to keep this in mind in considering evolutionary explanations of disease.

Roughly three overlapping research programs provide evolutionary perspectives on cancer: evolutionary medicine, evolutionary dynamics, and evolutionary developmental biology, or "evo-devo." First, evolutionary medicine is concerned with explaining the occurrence, prevalence, and natural history of diseases in light of evolutionary history (Gluckman et al., 2009; Stearns and Koella, 2008). This approach may draw upon anthropology, archaeology, genetics, and comparative biology. Investigating differences and similarities across different species and higher taxa may help make sense of why some species are particularly vulnerable to cancer or are more or less vulnerable to some types of cancer. Comparative biology may help us identify mechanisms associated with cancer vulnerability, onset, and progression. Some of the same mechanisms involved in the inhibition of cell proliferation, the regulation of cell death, resource transport, and the creation and maintenance of the extracellular environment are shared across species, and many of these are altered or co-opted in cancer (Aktipis et al., 2015). Comparing and contrasting how these mechanisms either act in the suppression of cancer-like growth or are co-opted in cancer progression, across different species, can help cancer researchers identify targets of treatment or prevention. One may also look deep into the evolutionary past to better understand how multicellular organisms evolved, and thus how cooperation among cells evolved. The mechanisms that make cooperation possible are often exactly those that prevent or slow the onset of cancer.

Second, one can use the tools of mathematical biology to model cancer's evolutionary dynamics, or patterns and processes of progression within a single tumor or across different cancer types or subtypes. Mathematical biologists have developed a variety of theoretical models of cancer progression. Some represent the population of cells in a tumor or leukemic cells as akin to an evolving population; others represent cancer progression as akin to competitive exclusion in ecology. One might model, for example, relative responses to chemotherapy in tumors of different sizes, with different rates of growth, death rates, or mutation rates (Wodarz and Komarova, 2014). Or genomic sequencing of tumor biopsies from living or deceased patients are used to model the evolutionary dynamics and progression of a cancer over time in a single patient, effectively mapping the "phylogeny" of a single cancer (Gerlinger et al., 2012; Navin et al., 2011; Yachida et al., 2010).

Third, evolutionary developmental biologists are concerned with how evolution shapes development and how development shapes evolution. This research program has much to teach us about how, when, and why cancer is more or less likely (Peirce et al., 1978; Reya et al., 2001; Gilbert and Eppel, 2009; Huang et al., 2009). Cancer has a variety of features that make it akin to the process of "arrest" in development. Cancer cells are often described as "de-differentiated": they lose many of the features typical of the cell of origin—a kind of regression to an ancestral state. Leukemias, for instance, are caused in part by an arrested development of blood cells in an early stage of development. Many cancers contain cells with "stemlike" properties (Reya et al., 2001). Understanding how and why these cells acquire this feature could help explain why cancers that have such properties tend to be more resilient and susceptible to recurrence than others (Laplane, 2016). Clearly, a better understanding of the mechanisms governing development, differentiation, and their evolution might better inform our understanding of cancer and enable better means of intervention.

Evolutionary explanations of disease, generally, have very different targets: disease persistence in the human population, differential incidence and mortality in distinct human populations or across different species or lineages, or the natural history of a particular disease in a particular individual. When debating the merits of evolutionary explanations of disease, we must be specific about the target explanandum; not all evolutionary explanations of disease are of the same target (persistence, differential vulnerability, or disease course). Evolutionary explanations of all of these explananda need not be in tension; indeed, they can be mutually informative. For instance, perhaps our evolutionary history left us vulnerable to cancer, which explains why this disease persists. There may, however, be differential vulnerability due to variability in mechanisms, features, or structures that make us more or less vulnerable to cancer. The course of a specific disease in a specific individual—perhaps an improvement with treatment, followed by treatment failure—may be best described in many cases in terms of adaptation of populations of cells in a tumor.

Figure 5.1 identifies distinct types of evolutionary explanations of disease persistence or presence in current human populations. This taxonomy is an elaboration upon the classification scheme of Murphy (2006), though I also draw upon Gluckman et al. (2009). There are some things worth noting about this taxonomy. The first is that evolutionary explanations of disease are not mutually exclusive. The same disease may exhibit features of "breakdown" and also be a byproduct of selective trade-offs. For instance, most of the diseases of old age (osteoarthritis, cancer) are due to a gradual breakdown or compromise of functional structures or processes. Cancer is (in part) a breakdown of the complex suite of

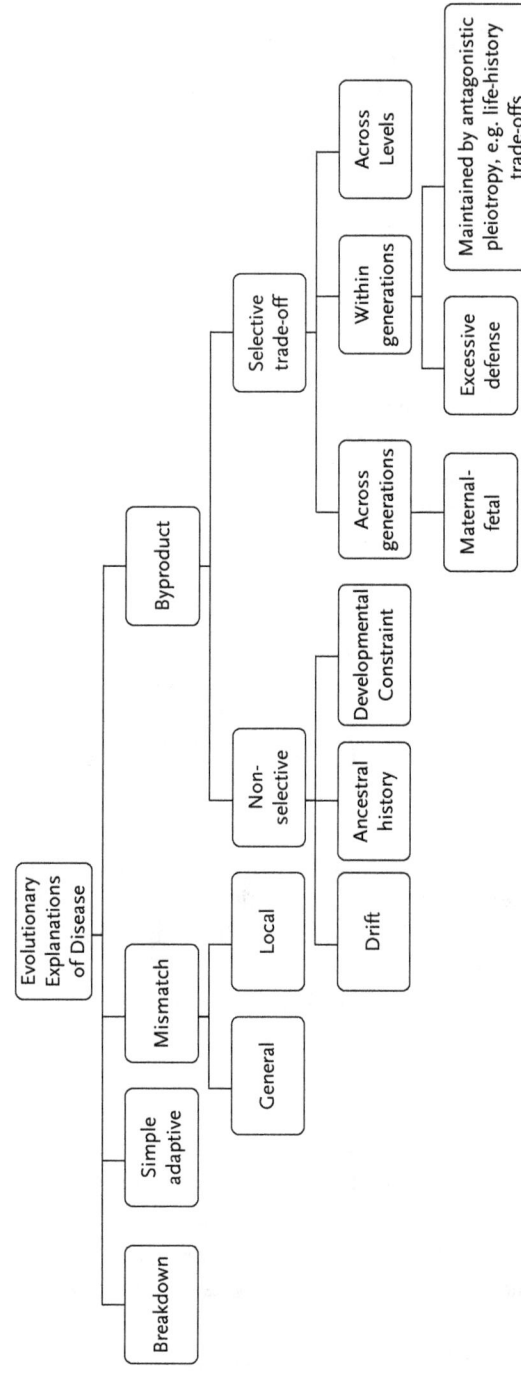

FIGURE 5.1 Types of evolutionary explanations of disease.

inter- and intracellular signals that control cell birth and death. Autoimmune diseases might also be considered a failure or breakdown of appropriate function of the immune system. However, autoimmune diseases may also be the product of an excessively sensitive or reactive immune system. So autoimmune diseases might be an "adaptive defense" that has gone awry, overreacting to inappropriate stimuli, though they are also a breakdown of function. These claims are not in tension, but mutually informative; whether a trait is "functional" is often a matter of timing and rate of expression of a capacity or disposition, not simply the presence or absence of a trait.

A second thing to notice is that the largest category in this taxonomy is "byproduct." Many diseases are in some sense an accidental byproduct of contingent facts about a lineage's history or ancestry; for example, a population bottleneck may have led to the fixation of a deleterious gene or genetic combination. This may explain many familial or strongly inherited diseases, such as hereditary forms of non-polyposis colon cancer in Amish populations (Orton et al., 2008). Diseases may also be byproducts of otherwise selectively advantageous traits. Changes in our environment or our lifestyle may leave us more vulnerable to diseases unknown in the evolutionary past. Selective trade-offs in life history or development leave us vulnerable to disease. Some sexes may be more or less vulnerable to specific cancers, and even within a sex, some groups may be more vulnerable than others. For instance, men's relative vulnerability to prostate cancer may be positively correlated with CAG-repeat numbers. The number of CAG repeats is responsible for levels of androgenic hormones, which in turn can increase physical height and rate of growth. Some argue that this trait may have been selectively advantageous in early life but may yield a trade-off in fitness later in life, as it could lead to higher rates of prostate cancer (Summers and Crespi, 2008).

Some disease symptoms (and causes) are an adaptive defense against other diseases or potential environmental harms. For instance, fever and mucus production are reactions to infection, as is inflammation; both promote faster healing (Gluckman et al., 2009). However, inflammation (especially persistent inflammation associated with infection) increases risk of cancer. A response to infection could itself become so extreme as to be classified as "disease"; excessive fever, leading to seizures, may be due to overactive defense mechanisms.

One can also seek to explain increases in disease incidence by appealing to evolutionary considerations. Some argue that increases in cancer incidence in modern industrial societies are due in large part to rapid shifts in our environment and life history (or "rapid" in evolutionary terms). On this view (Greaves, 2001; Stearns and Koella, 2008), changes in human lifestyles have altered our endogenous and exogenous environments in ways that predictably yield higher rates of cancer. We live longer, delay reproduction, consume unhealthful foods in greater excess, are exposed to a wide variety of environmental carcinogens, and have sedentary lifestyles, in comparison with our ancestors. All of these changes to our lifestyles place stress on various systems of repair and regulation of cell birth and death. These changes affect our metabolism and microbiome, and both such changes can lead to obesity, inflammation, excessive immune response, and perhaps changes to the regulation of our endocrine system, which in turn may increase our risk of cancer or simply permit greater opportunity for cancer. Such explanations treat our relatively high incidence of cancer as due to a "mismatch" between our current and ancestral environments.

According to Greaves (2001), hormonally driven cancers—estrogen-receptor-positive (or ER-positive) breast cancer in particular—are in part a product of a mismatch between our ancestral and current lifestyles. He argues that avoiding or delaying reproduction exposes women to excess cycles of estrogen, which increases risk of ER-positive cancers (Greaves, 2001). Women in the developed world, who often delay (or refrain from) conceiving, tend to exhibit higher rates of ER-positive breast cancer than women in the developing world. That is, women in the developing world are more likely to have more children and to start having them at a younger age, which was far more common in our evolutionary past. Moreover, early onset of menses in the relatively well-fed developed world is also associated with higher rates of ER-positive breast cancer. Defenders of evolutionary medicine have argued that these disparities are best explained by the "mismatch" between our current and past reproductive practices, as well as mismatches in our modern diet with that of the evolutionary past. According to Greaves (2001), in our ancestral environments, spending the majority of their reproductive years pregnant or nursing protected women against ER-positive breast cancers.

Of course, it's possible that other factors may better explain this correlation (e.g., exposure to estrogenic elements in our current environment, particularly early in development; for a discussion, see Cranor, 2011). Moreover, it is highly likely that breast cancers are not discovered as frequently or documented as thoroughly in the developing world as in the developed world; so higher incidence may just be an artifact of more frequent screening. This case is also complicated by the fact that there is a slight increase in risk of cancer immediately following pregnancy, falling off over time. The risk is slightly more elevated and falls off more slowly for those who delay pregnancy longer (Schedin, 2006), which is consistent with the evolutionary explanation given earlier but raises independent questions about the value of pregnancy in protecting against breast cancer. These "pregnancy-related" cancers are also likely to be associated with hormonal changes during pregnancy, as well as dramatic changes in the tissue microenvironment of the breast during and after pregnancy and nursing. The involution of milk ducts after nursing involves cellular changes akin to wound healing, and so possibly increased chances of dysplastic growth, as cancer progression co-opts many of the signals associated with wound healing (Weinberg, 2007).

In sum, while evolutionary explanations can be tempting, there are always plausible alternative explanations. Determining how much of the differential vulnerability to breast cancer in the developed world is due to low parity or early menses, and how much to environmental exposures, is not straightforward. Moreover, explanations that make reference to our purported ancestral environments are vulnerable to a number of criticisms (Murphy, 2006). First, of course, one needs to establish what was common or rare in our ancestral environments, as well as what is most likely to have been adaptive in such environments (see, e.g., Buller, 2005). The assumption that there was one relatively stable such environment (the "environment of evolutionary adaptedness") is hugely contested; there was likely to have been variation in environments to which humans adapted over their evolutionary history, and, of course, humans have shaped their environments in ways that have had a significant impact on their selective history. Second, heritable variation is necessary for selection to act on a trait; for selection can act only when there is heritable *variation* in fitness. But establishing any of these claims requires extensive evidence, and this evidence is difficult to gather and notorious for underdetermining competing hypotheses. Extrapolations from such evidence are enormously difficult to support, and sensitive to initial assumptions. Perhaps needless to say, evolutionary explanations of complex behavioral traits are plagued by similar challenges (see, e.g., Buller, 2005; Downes, 2008; and, in the context of psychiatry, Murphy, 2006).

Nonetheless, there are several relatively uncontroversial ways in which our evolutionary and developmental history helps to predict and explain various aspects of cancer. First, and perhaps most obviously, only multicellular organisms like us are subject to cancer. Multicellular individuals are cooperative groups of cells. The emergence of mutlicellularity required the evolution of a variety of mechanisms for the enforcement of cooperation among these cells (cf. Queller and Strassman, 2009, 3143; see also Buss, 1987; Maynard Smith and Szathmary 1995; Michod 2000). Cooperation is not, of course, an inevitability; interactions in biological collectives can result in mutualism, cooperative organization, parasitism, or mutual destruction. Cancer is an instance of the latter; indeed, cancer is (in part) a breakdown of the mechanisms that ordinarily enforce cells' cooperation. Godfrey-Smith (2009a) calls this "re-Darwinization," a return of our component cells to a relatively autonomous state. Cancer is one of many examples of how mechanisms that enforce cooperation and suppress conflict within biological collectives can fail.

Multicellular organisms like us have distinctive functional organization and a high degree of plasticity, redundancy, and modularity. These features are adaptive; redundancy enables functional parts to take over when one such part fails, and modularity (some claim) enables "evolvability" (the capacity to evolve, though see, e.g., Pigliucci, 2008, and Brown (2013) for discussion of the variety of senses of "evolvable"). That is, some argue that modular lineages can generate more evolutionary novelties, because parts may vary independently without compromising the fitness of the organism as a whole (see, e.g., Simon, 1962; Wagner, 2013). However, some of these features also make us vulnerable to cancer. The plasticity of our cells, in particular the capacity to transition from epithelial to mesenchymal cell types, is unique to metazoans and in part explains how metastasis is possible in carcinoma. Without this capacity, metastatic cells could not escape the epithelium and enter the blood and lymph system (Thiery, 2002). Greater or lesser vulnerability to cancer is also lineage-specific. Whereas humans have only one copy of *TP53* (a gene associated with tumor suppression), elephants have twenty copies, a situation that may have evolved to serve a protective function, given elephants' relative size and long life span (Abegglen et al., 2015). That is, larger body size and longer life span are associated with more somatic cell divisions, and so more mutations, and thus higher rates of cancer. This would create selective pressure for more tumor-suppressive genes.

In sum, there are (potentially) many evolutionary explanations of cancer's origins, dynamics, and differential vulnerability, some that appeal to general features of lineages or particular species, some that appeal to evolutionary dynamics in populations, some that appeal to shared mechanisms, and some that appeal to trade-offs in fitness. In the next section, we will focus more narrowly on the role of mathematical and dynamical models in investigating and explaining various aspects of cancer initiation and progression and consider in what sense cancer progression itself is a process of natural selection. Taking a multilevel perspective on cancer—that is, viewing cancer progression as both a process and by-product of selection at several levels of biological organization—suggests new hypotheses to test and predicts, as well as explains, various features of cancer progression, including variation in cancer's natural history across individuals (e.g., variation in response to therapy) and across species.

5.2. Cancer Evolving

The idea that cancer progression may be viewed as an evolutionary process is not new: Gause (1966) compared a population of cancer cells to an evolving population of microbes.[1] Nowell (1976) and Cairns (1975, 1978) similarly argued that "cancer can be viewed as the operation of Darwinian selection among competing

---

1. This section is coauthored in part with Chris Lean.

populations of dividing cells" (1978, 151).[2] What started as a minority view has now become an active area of research, drawing upon different domains of biology (molecular genetics, phylogenetics, and population genetics) and different tools and technologies (mathematical modeling, computer simulation, and experimental evolution in model systems). A variety of mathematical biologists have modeled cancer progression as a process of natural selection (Michor et al., 2004; Merlo et al., 2006; Anderson et al., 2006; Pepper et al., 2007; Greaves and Maley, 2012). Such models can predict which cancers are more likely to evolve chemotherapy resistance under what treatment regimens (Komarova and Wodarz, 2005). Some agent-based models and simulations have been used to investigate which mutations in which order are likely to yield the most aggressive forms of the disease (Spencer et al., 2006).

The central idea behind this approach is as follows. First, cancer cells are cells that have acquired a series of somatic mutations and epigenetic alterations that allow them to escape regulation of cell birth and death, leading to disorderly growth, invasion, and metastasis. Over the course of the average human's lifetime, there are many millions of cell divisions in the body, and so, by chance alone, mutations and epigenetic changes occur. Some such mutations are associated with cancer: they may lead to chromosomal instability, failures of DNA repair, or failures in the regulation of cell birth and death, or the acquisition of cancer's "hallmarks."

There is, in fact, a "vast reservoir" of mutations in healthy normal cells, but most such cells do not eventuate in cancer.[3] How and when cells are born and die—that is, the particular mode of regulation of growth—is very specific to the type of tissue and the stage of development of the organism. Epithelial cells in the skin or colon, for instance, are born, die, and are replaced thousands of times over the course of a lifetime; in contrast, bone growth and renewal are relatively slow post-adolescence. Some such cells become invasive cancers. When and how they do has to do in part with cell-intrinsic features, but it also has to do with signals from the local microenvironment, immune response, and tissue architecture. Provided mutations to cancer cells (or proto-cancer cells) are heritable and make a difference to the relative success of cells and cell lineages, selection is acting on such populations. That is, the relative survival and reproductive success of cancer cells is, in principle, a product of natural selection, however short-lived.

However, selection does not occur only between cells. Cancer is also meaningfully described as a *byproduct* of selection at other levels of organization.

---

2. For a history of evolutionary perspectives on cancer, see Morange (2012).

3. As discussed in Chapter 1, according to one recent study, 18–32% of normal skin cells in the average sun-exposed adult have clonal populations of cells with two to three "driver" mutations (at a density of about 140 driver mutations per square centimeter) (see Martincorena et al., 2015).

Cancer cells co-opt signaling systems that ordinarily promote the survival and reproduction of the organism as a whole. A multilevel evolutionary perspective can thus shed light on both the dynamics of cancer progression and differential vulnerability to cancer within and between species and higher clades.

What is a multilevel perspective? Multilevel selection theory starts with the relatively uncontroversial observation that populations of entities are nested within one another in the biological hierarchy: populations of genes are nested within cells, which are nested within organisms, and organisms are nested within populations, which are nested within species, and so on to whole lineages. Provided that these entities vary, that the variation is heritable, and makes a difference to relative fitness, selection can act at more than one level in this hierarchy; for, in principle, any population of entities with heritable variation in fitness may evolve by natural selection (Lewontin, 1970). In a "multilevel" selection situation, selection is acting on more than one level simultaneously. Selection at one level may increase or decrease the frequency of traits in a population, which in turn may affect what is available to selection at another level of analysis. There is an extensive cannon of interdisciplinary work on multilevel selection (Lewontin, 1970; Wilson, 1975; Damuth and Heisler, 1988; Michod, 1997; Sober and Wilson, 1999; Okasha, 2005, 2006; Godfrey-Smith, 2009a).

In multilevel selection, it can become difficult to discriminate which outcomes are due to selection at a "lower," and which at a "higher," level of analysis. Damuth and Heisler introduced a useful distinction between multilevel selection 1 (MLS1) and multilevel selection 2 (MLS2) to assist in discriminating between different scenarios: "of interest in the former case [MLS1] are the effects of group membership on individual fitnesses, and in the latter [MLS2] the tendencies for the groups themselves to go extinct or to found new groups (i.e., group fitnesses)" (1988, 407). That is, in MLS1, fitnesses are properties of individuals, and character values are attributed to individuals. But since group membership can have effects on individual fitness, when individuals congregate in groups one may have a case of MLS1. For instance, when different groups have different mixes of individuals, and some mixes are more successful than others (i.e., the individuals in some groups better survive and reproduce than the same individuals would in a different group with different proportions of variants), we may describe outcomes of such group-mediated effects as a product of "group selection" in the sense of MLS1. The outcome considered in such cases is the changing proportion of kinds of individuals in the whole population (the metapopulation). Damuth and Heisler explain that some prefer to call MLS1 "group selection" because group-level properties (such as the relative proportion of altruists in a group) determine the relative success of these individuals.

There is another sense of "group" selection that is often confused with this first sense: what Damuth and Heisler prefer to call multilevel selection 2 (MLS2). Here "group selection" refers to change in the frequencies of different kinds of groups in light of "group-level" properties. In MLS2, character values are attributed to groups (including both aggregate and global characters), and it is these group-level characters that make a difference to group success at propagation, or the seeding of more groups. Groups are more "fit" if and only if they propagate or seed more groups. To be clear, it is not a necessary condition of MLS2, according to Damuth and Heisler, that group characters be "emergent" in the strong sense that there is no way to explain the properties of the group in terms of the properties of the individuals. *Average* height or extent of genetic heterogeneity in a population is, after all, a *group-level* trait, but it can be explained in terms of the properties of individuals that make up the group. All that MLS2 requires is that these group-level properties are what cause group success (measured in this case by the seeding of new groups) (Damuth and Heisler, 1988). As a vivid but simple example, consider two groups of organisms: one that is genetically uniform and one with a large amount of genetic variation. In the latter group, greater extent of genetic variation may better enable such groups to propagate and diversify. Insofar as this group-level trait (extent of genetic heterogeneity) was a cause of relative group success at propagation, this would count as an instance of MLS2.

Multilevel selection theory has been used to explain the emergence of new levels of organization in the biological hierarchy, sometimes called the "major transitions in evolution" (Okasha, 2005, 2006). The evolution of multicellularity required the emergence of cooperative interactions between populations of cells. There were intermediate stages between functionally integrated individuals and mere collectives. This may have involved the co-option of traits initially adaptive at the individual level in service of collective benefit. Quorum sensing in bacteria may be a vivid example. Quorum sensing is a method by which bacteria detect local cell density by the diffusion and accumulation of signaling molecules in the surrounding environment. At certain densities, the population as a whole can engage in a coordinated response—for example, the production of bioluminescence, population mobility, biofilm maturation, and virulence (Williams et al., 2007). That is, quorum sensing regulates the production of "public goods," or products manufactured by an individual that can then be utilized by the individual or its neighbors. The capacity to quorum-sense likely evolved from "self-signaling" systems that were adaptive at the individual level (Redfield, 2002; Williams et al., 2007). There are many such cases of proto-organismic collectives in nature. Or as Queller and Strassman put it, there are

> levels of organism, and each level was attained by merging formerly separate individuals at a lower level . . . Multi-cellular individuals are cooperative groups of cells, eukaryotic cells are cooperative assemblages of multiple prokaryotic lineages and prokaryotic cells must have emerged by

assembly of formerly independent replicators. These major transitions in evolution construct new levels of organism out of separate individuals. (2009, 3143)

How do these transitions come about? Multilevel selection theory figures in one prominent account: During the transition to multicellularity, the fitness of the collective must override the fitness interests of the parts. Collectives or groups must become more successful *as collectives*; in other words, they have to outcompete other collectives in reproduction or in the propagation of further collectives. Cooperation and functional integration of parts can play a role in the evolution of such a collective. Co-option might also play a role in the emergence of functional integration. In the case of quorum sensing, traits that originally may have enhanced individual fitness can be co-opted in service of collective fitness, or the relative success of groups at propagation. So what might begin as a "group selection" process in the sense of MLS1 (individuals benefiting due to group membership) could evolve into a circumstance where we have a case of MLS2 (groups being more or less successful at propagation due to group-level properties) (Michod et al., 2006; Okasha, 2006).

Whether or not MLS1 or MLS2 played a role in the emergence of multicellularity, it's relatively uncontroversial that the major transitions in evolution involved successive changes in population structure and organization. Individuals gradually organized into collectives and, eventually, came to succeed or fail as collectives. If particle fitness is improved by particles being part of a particular collective, then the particles may be spoken of as having "aligned" fitness. Aligned fitness allows for the emergence of complex integration and interaction within the collective, as well as division of labor, such that the collective fitness of the group can be spoken of as a whole.

To summarize, in principle, selection can act within a collective—among the parts—and between collectives. These processes could be going on simultaneously or sequentially. Damuth and Heisler (1988) describe two senses of multilevel selection, MLS 1 versus MLS 2, to highlight the fact that "group selection" may refer to different things and that the collective fitness of a group is not simply a matter of averaging over particle fitnesses (see Table 5.1 for a summary). Critics of genic selection (e.g., Sober and Wilson, 1999) have emphasized this important difference: simply averaging the fitness of different genotypes that live in distinct groups may result in a loss of important causal information, or it may erase the distinction between these two selective processes. The fitness of a group is not simply the average fitness of its members: the two can come apart and pull in different directions. Some groups may propagate more successfully than other groups, even if the average fitness of the group's members is low. For instance, as Wilson demonstrated

Table 5.1 The Distinction between Multilevel Selection 1 (MLS1) and Multilevel Selection 2 (MLS2) According to Damuth and Heisler (1988)

Multilevel Selection	MLS1	MLS2
"Group selection" refers to:	The effects of group membership on individual fitness	Change in frequencies of different kinds of groups
Fitnesses are properties of:	Individuals	Groups
Characters are values attributed to:	Individuals (e.g., altruism)	Groups (e.g., group mean, population density, proportion of different phenotypes)
Appropriate for investigation of:	Evolution of characters of individuals likely to be affected by group membership	Changing proportions of different types of groups; different propensities to go extinct or to found new groups

(1979), group membership can make a difference to relative fitness; what he calls "trait groups" are groups where population structure is not freely mixing but where similar individuals tend to find themselves together. Cooperation is more likely to evolve in such a situation than in a freely mixing population, where "cheaters" or "free-riders" can arise and swamp the population as a whole. In trait groups, altruists might benefit from group membership, even though altruists by definition have lower fitness than selfish individuals in the population at large.

How does this picture apply to cancer? Models of cancer's evolutionary dynamics represent either cells or whole populations of such cells as subject to selection; some models focus on interactions between cancer cells and their microenvironment or the particular features of subpopulations and their causal interactions with their local microenvironment (Wodarz and Komarova, 2014). Taking a multilevel evolutionary perspective can assist in integrating a variety of levels of analysis or show how they are mutually constraining and mutually informative. There are potentially several different targets or levels of selection at play in cancer initiation and progression, some more salient at different stages of progression.

Identifying a level of selection involves identifying a population of like entities, which have reproduction and differential fitness due to heritable features of such entities (Godfrey-Smith, 2009b). According to the multilevel selection view, there are several distinct units or targets of selection that are potentially of relevance in cancer progression. These are not mutually exclusive options, but may be co-occurring:

1. Individual selection, mutation, and drift, acting among single cells, whether "pre-cancer" cells, cancer cells located in a tumor, or metastatic populations of cells entering the bloodstream. Individual cells may be more or less fit in the sense of overcoming the variety of constraints in normal tissue that ordinarily prevent disorderly cell growth, acquiring resources in the local tissue micro-environment, or adapting to novel tissue microenvironments (in the case of tumor metastases).

2. Group selection in the sense of MLS1 or MLS2. Some individuals may survive and reproduce more successfully than others, because of their membership in a group with a distinctive distribution of variants (MLS1), or relative fitness may be ascribed to groups as a whole, provided features of the population as a whole make a difference to the relative "reproductive" success of the group—or the capacity of a group to seed new groups (MLS2).

How are MLS1 and MLS2 relevant to cancer progression? Stepping back, one might view the process of cancer progression as proceeding through a series of stages, where simple selection is more salient in the beginning, and gradually MLS1 and MLS2 may become salient as well (see Figure 5.2), not unlike the process some have hypothesized was at work in the emergence of multicellularity, as discussed earlier (see, e.g., Okasha, 2006). Most models of the evolutionary dynamics of cancer are simple individual-level selection models (Nowak, 2006; Frank and Nowak, 2004; Michor et al., 2004), where fitness is attributed to a single cell due to a gain of function or loss of function: promoting angiogenesis, blocking tumor suppression or inhibitors of growth (halting or locking apoptosis, or cell death), or promoting cell division. Growth in any environment will be limited by space and the availability of nutrients; among incipient cancer cells, some cells will be better able to compete for these resources or better able to escape mechanisms of homeostatic control. Such evolutionary dynamics are often taken to feature centrally in early stages of the development of a "proto-cancer cell" lineage. Models of such circumstances look not unlike competitive exclusion models in ecology.

There could also be selection operating at the level of cell colonies, however. Komarova and Wodarz (2004) argue that there is an optimal "strategy" for chromosomal instability (CIN) or rate of loss of heterozygosity in cancer. Cancer cells characteristically have various chromosomal abnormalities—duplications, losses, and inversions of large parts of the genome complement of a cell. One might

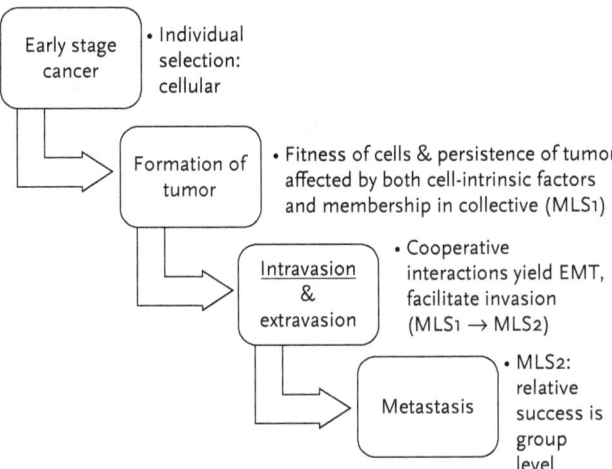

FIGURE 5.2 One model of how multilevel selection may play a role in cancer progression.

think CIN is uniformly bad, but in cancer it is a two-edged sword: too much instability can lead to destructive genomic changes, such as the loss or destruction of whole chromosomes, but if timed correctly, CIN can inactivate tumor suppressor genes and thus promote clonal expansion. Some lineages will thus be more successful than others in progression to metastasis due to the specific types of traits of *cells*, traits acquired at specific times or stages in the progression of the disease. But some subpopulations of such cells may more quickly reach a cancerous state, insofar as they have the optimal "strategy"—or optimize the rate of growth given a certain degree of instability. Too much instability is detrimental because it increases the death rate; too little slows down progress because the rate of onset of cancerous mutations is low. Of course, this is an idealized model, and populations of pre-cancerous cells are not a simple series of clonal expansions, acquiring a specific mutation at each stage and not "adjusting" their rate "strategically." But Komarova and Wodarz's point is a theoretical one: early acquisition of CIN carries an advantage, which, they argue, falls off over time, as too much genetic instability overall reduces fitness; lineages with the optimal strategy are the ones that succeed in progressing to cancer.

Selection might also be going on simultaneously at multiple levels of analysis in the evolution of chemotherapy resistance. The relative success of cancer cells at evolving resistance to chemotherapy depends upon cells acquiring drug resistance, but evolutionary principles can predict when and why combination therapy will prevent treatment failure in populations of varying sizes, with varying mutation rates, or with higher or lower rates of turnover. For instance, for a given population size, a higher mutation rate will result in a higher rate of evolution of chemotherapy resistance (Komarova and Wodarz, 2005). One may also predict that combination therapy will be by and large less successful in populations with a higher rate of turnover of cells. The general principle is the same as that used to explain why bacteria evolve antibiotic resistance more quickly than humans evolve resistance to bacteria.

Cancers also may be more or less successful at metastasizing, and some cancers are more successful at metastasizing to different locales, given different "pre-adaptive" features of the original cells. What has been called the "seed and soil" hypothesis is that some cancer types are more likely to metastasize to some sites in the body and not others. For example, breast cancer is most likely to metastasize to the lungs and brain. This is likely due to distinctive features of these cells, but also to systemic features of the metastatic sites. That is, the "soil" is as relevant to cancer progression as the "seed"—cancers may be spoken of as having "an ecology"—features of the surrounding tissue that select for variable features in cancer cells themselves. This is not a new idea, but one that lends itself handily to the evolutionary picture of cancer (Fidler, 2003). That is, the environment itself plays a role in cancer progression, shaping the cellular phenotype, and it appears that cancer in turn shapes its tissue microenvironment (Egeblad et al. 2010). Committing to the view that cancer progression is akin to a process of selection need not commit one to naive genetic determinism or reductionism; evolution is a process of interaction between organisms and environments (Levins and Lewontin, 1982).

Moreover, population structure (or the "architecture" of a population) could well predict the evolution of faster or slower progression to cancer phenotype. The time to the evolution of escape from feedback regulation typical of normal tissue depends upon effective population size, which is a group-level property (MLS2). The "cancer stem cell" hypothesis holds that some cells in a tumor have stemlike properties. Such cells are constantly renewing and thus more likely to both seed new tumors and perhaps also survive chemotherapy (Reya et al., 2001). While the cancer stem cell hypothesis is somewhat controversial (see, e.g., Gupta, 2009; Nguyen et al., 2012), it is certainly true that some cells in a cancer appear to be better able to seed new tumors. In this case, a population of proto-cancerous cells with a larger number of stemlike cells (or cells with a distinct capacity to escape regulation from the tumor microenvironment) increases the relative "fitness" (in the sense of rapid growth) of the group.

Critics of the Darwinian picture have argued that models of cancer evolution treat cancer as a simple clonal expansion process, where the population of cells is relatively homogeneous (see, e.g., Weinberg, 2013, 458–468). However, there is no reason in principle why population structure in a tumor cannot be incorporated into an evolutionary model (indeed, it's been done already; see, e.g., Wodarz and Komarova, 2014). As Weinberg (2013, 464) points out, tumors may in fact contain multiple coevolving subclonal lineages arising from distinct stem cells. Such coevolving subpopulations—or the coevolution of metapopulations—are fairly common in nature and have been studied in a variety of species where lineages are subdivided into local niches and so may either evolve local adaptations or engage in competition or coevolutionary dynamics. Weinberg considers whether "distinct subpopulations of cells within a tumor act symbiotically, each supporting the growth and survival of the other, rather than directly competing" (466). Such coevolving lineages would count as a case of a higher level of selection; for example, such heterogeneous tumors may be more successful at acquiring resources, resisting attacks from the immune system, or seeding metastases, which effectively counts as an instance of MLS2.

MLS2 becomes more relevant as we consider malignancy, or invasion and metastasis. This process requires that the carcinoma grow to a stage where it can acquire a novel phenotype (motility) and thus reproduce: metastasis is the production of other tumors throughout the body. But this process is not simply the release of a single cell into the blood or lymphatic system. "Seeding" of metastases is an extended process that involves a complex series of steps: the breaching of the basement membrane, intravasion, the invasion or dissemination of metastatic cells, and the colonization of distant sites. This process requires the cooperation of tumor cells and stroma, the co-option of the very same heterotypic interactions that are ordinarily used in service of wound healing (see, e.g., Weinberg, 2013, 587–607). Most attempts at seeding or most incipient metastases fail; the disseminated cells more often than not are destroyed in circulation. The ability of micrometastases to colonize remote sites is not simply a given, in other words, but requires many attempts, akin to the colonization of remote islands by Darwin's finches. Some metastatic populations are more effective at colonization than others, and once a metastasis is established, these colonies, in turn, can be more or less successful at disseminating secondary metastases. The secondary wave of metastasis is sometimes called a "metastatic shower" because tumors send off many "showers" of potential colonists (cf. Weinberg 2013, 654).

The success of metastases is not due exclusively to properties of metastatic cells. Metastasis involves the cooperation of the stromal tissue; activation of the EMT (the epithelial–mesenchymal transition) in carcinoma requires signals from the tissue microenvironment; and the tumor both recruits stromal cells (macrophages, mast cells, fibroblasts) and co-opts signals ordinarily associated with wound healing to break down surrounding tissue and metastasize. That is, the entire process of metastasis is not achieved by single cells in isolation, but by the activity of both the tumor and tumor-associated cells in stimulating angiogenesis, breaking down surrounding tissue, and enabling intravasion and invasion. We might think about metastasis as rather like "r" selection rather than "k" selection;[4] given that the environment is unstable, and most metastases fail, tumors more successful at sending out large populations of multiple metastases are more likely to be successful. More heterogeneous primary tumors—in the sense of both more genetically diverse and more epigenetically plastic and modular— may be able to generate more and more successful metastases. And evolutionary dynamics may also apply outside the primary tumor; metastatic populations may be more or less successful at colonizing and may be more or less likely to generate more and more successful colonists (or secondary metastases). There is in fact evidence that the ability to colonize evolves (see Klein et al., 2002). Differential success in this case is due in part to group-level properties, namely genetic heterogeneity and functional diversity. Different cell lines in a tumor might acquire differential capacities that, combined, lead to more successful reproduction of secondary metastases.

Thus, depending on the level of analysis and the empirical details of any particular case, cancer progression could be viewed as an individual-level selection process or a group-level selection process, in the sense of MLS1 or MLS2. That is, a cancer cell could be competing with or coevolving with other cell lineages, or incipient cancers or proto-cancer cell populations might be competing with other proto-cancers for space and resources. Or differential reproduction via differential success at invasion and metastases could be due to complex properties of the tumor as a whole, including functional differentiation within the tumor and co-option of organismic adaptations in service of invasion and metastasis. Wodarz and Komarova make the point quite simply:

> Assume that tumor progression requires the presence of three gene products, call them A, B and C. A cell can evolve to accumulate all three mutations, thus achieving the malignant phenotype. On the other hand, the three gene products can arise independently in different cells, and they can share the gene product as a public good. In this setting, the malignant phenotype arises as an emergent property from a collection of cooperating cells rather than an individual cell. (2014, 460–461)

Put simply, on the MLS2 view, collectives are more or less successful in light of group-level properties: heterogeneity, or division of labor in the process of invasion and metastasis. In the latter case, the collective of cell lineages in the tumor

---

4. "r" and "k" refer to growth rate and carrying capacity in standard ecological theory. So, "r-selected" species are those that produce many offspring, each of which has a relatively low probability of surviving to adulthood (high "r", low "K"). "K-selected" species are those that have lower growth rates, and tend to have higher survival rates. Typical examples of "r-selected" species are dandelions; elephants are "K-selected."

in some sense has "aligned" fitnesses. Reproductive "success" in such cases is measured in terms of the invasion and propagation of metastases (recall that most incipient cancers fail to progress); or metastases could be more or less successful at generating secondary metastases. In MLS2, we are interested in the way group-level traits affect group-level fitness. Cell lineages (including cancer and non-cancer stromal cells) might interact as a collective (both with other cancer cells and with the tumor stroma), thus increasing the collective's ability to survive and reproduce relative to other collectives (which might include other pre-cancerous lesions or other metastases at seeding further "secondary" metastases). If better functional integration in a tumor results in more successful metastasis, then the collective fitness of the whole is distinct from the aggregative fitness of the individual cells. There is now ample evidence that tumors are complex heterogeneous populations of cells and that signaling between these cells plays an essential role in tumor progression (Egeblad et al., 2010).

Cancer cells also recruit neighboring cells in order to participate in tumor development, both as a structural support and in building a blood supply (Mueller and Fuesnig, 2004). Arguably, this is a form of *niche construction* in which the cancer cells and normally functioning cellular populations both evolve and coevolve with their surrounding environments (Odling-Smee et al., 2003). That is, cancer progression requires the cooperation of the tissue microenvironment; the behavior of a tumor is not due exclusively to the behavior of cancer cells alone. Carcinoma cells release growth factors that recruit macrophages, neutrophils, and lymphocytes, which in turn orchestrate an inflammatory response that involves the release of signals that stimulate the proliferation of epithelial cells and the process of angiogenesis in the tumor stroma. Epithelial cells release platelet-derived growth factory (PDGF), for which fibroblasts and myofibroblasts have receptors; these in turn release IGF-1, which benefits the growth and survival of nearby cancer cells. Stromal and epithelial cells collaborate in the construction of the extracellular matrix (ECM). Stromal cells are recruited by tumor cells to aid in the construction of tumor colonies. Metastatic cells can spread throughout the body because they can acquire a "disguise" of platelets that enable them to evade detection (Egeblad et al., 2010).

In all these cases, cancer cells exploit preexisting biological programs that are normally used by tissues for other purposes, in particular wound healing. Tumor cells activate a complex program that is already available. Wound healing uses a number of growth factors to increase the permeability of the blood vessels near the wound, attract fibrogen to create blood clots, and attract fibroblasts, converting them into myofibroblasts, which in turn release matrix metalloproteinases (MMPs), which play a role in building up or remodeling the ECM, carving out spaces for new cells. Recruiting fibroblasts to the tumor allows the creation of a fibrin matrix, which serves as scaffolding in the initial formation of a tumor.

It's important to see that this cooperative process is not in tension with viewing tumor progression as a process of natural selection; selection and niche construction could be going on simultaneously. Moreover, populations of tumor cells more effective at shaping their environment to suit their ends will outcompete their peers.

The cooperation of the tumor cells and stromal cells is not simply a co-option of an existing program of cells, however. The cancer cells and non-cancer cells may *coevolve*; that is, cancer-associated stromal cells become more adept over time at assisting in the survival of cancer cells within the tumor. Experimental evidence for the importance of the stromal tissue in tumor progression comes from a variety of sources, but we will discuss just one powerful case study. Cancer-associated fibroblasts (CAFs) are very similar to fibroblasts in normal cellular tissue, which play a role in building up scaffolding in injured tissue. In an experiment to test the role of CAFs, normal human prostate cancer cells were mixed with different kinds of fibroblastic cells and then implanted in host immunocompromised nude mice. When cancer cells were mixed with CAFs, they grew tumors five hundred times as large as when the CAFs were mixed with fibroblasts from normal prostate glands prior to injection into hosts (Olumi et al., 1999). Weinberg argues that the cumulative evidence of the role of CAFs in tumor formation suggests that "stromal cells in advanced carcinomas, having coexisted with epithelial cancer cells for many years, may change their genotype and acquire traits that genetically normal stromal cells cannot achieve. This suggests that stromal cells co-evolve with their neoplastic neighbors during long periods of tumor development" (2013, 603). In other words, a tumor (which is composed not only of cancer cells but also of stromal tissue) is a complex tissue with coevolving cell lineages, composed not only of cancer cells but also of non-cancer cells, which together cooperate in the progression of a tumor. Egeblad et al. (2010) claim that the highly integrative functional organization of some tumors suggests that they should be considered to be akin to organs.

While there are multiple levels of selection in operation that potentially play an explanatory role in cancer, there must be a role for drift. There is no doubt that the population size of cells in a tumor is sufficiently large for selection to overcome drift; the volume of a cubic centimeter within a tumor cell mass contains about 10^9 cells, but tumors—to say the least—have a complicated population structure. It is quite likely that whether a cancer cell lives or dies may have to do as much with accidents of location as particular fitness-enhancing features intrinsic to the cell. The relative importance of selection versus drift will vary case by case; for instance, there are likely many isolated sublines of cancer cells within a tumor that may or may not survive, given lack of access to a blood supply or architectural features of the tissue microenvironment, rather than features intrinsic to the cancer cells or populations of such cells (Crespi and Summers, 2006). The threshold of 0.2 mm is the maximum permissible for oxygen to effectively diffuse between tissues; so tumors will not grow if they are at a greater than 0.2 mm distance from blood vessels. This mimics the way the fitness of normally functioning somatic cells is determined by their local cellular signaling environment rather than their own intrinsic features. Further, where a cancer is initiated may affect how slowly or quickly it progresses. Drift may well explain a good deal of the variation we find in whether and why tumors progress to metastasis.

In sum, this multilevel perspective provides a representation of cancer progression that is both predictive and explanatory. The accumulation of mutations (and epigenetic changes) in neoplastic progression involves heritable variation in the fitness of cancer cells and lineages—for example, the timing and acquisition of particular mutation types, chromosomal instability, hypoxia, acid resistance, and transition to mesenchymal phenotype. Cancer progression may involve several transitions in individuality and levels of selection and is a unique case of multilevel selection: it evolves at the level of populations of cells, via competition among cell lineages, via co-option, via a kind of cross-level exaptation, and as a group or higher-level "individual."

That is, the progression of a tumor may be viewed as the emergence of a new kind of evolutionary individual. Admittedly, the evolutionary process that this individual participates in is short-lived, but some advanced tumors have relatively complex functional organization, namely a division of labor where some cells function as the "sex" cells, showering the body with multiple metastases, and others as the "somatic" tissue. Some tumors are clearly more successful than others at evading chemotherapy and metastasizing, and arguably taking an evolutionary perspective can explain this success: highly genetically heterogeneous tumors are more likely to resist chemotherapy than less heterogeneous tumors.

In sum, evolutionary thinking about cancer can be an enormously fruitful tool: for generating novel testable hypotheses, for explaining the dynamics we do see, and even for generating novel frameworks or models for understanding cancer's etiology. Taking a multilevel perspective on cancer suggests that there may be multiple targets or units of selection evolving and coevolving in a cancer, even as cancer itself can be viewed as a byproduct of selection acting at the level of individuals and indeed whole lineages.

5.3. Evolution, Modeling, and Hypothetical Explanation

As we've seen, mathematical models of cancer's dynamics vary in terms of their explanatory targets—or the questions they are used to investigate. Their assumptions thus vary, and so also does their representativeness of various aspects of the complex process of tumorigenesis. Mathematical models are deliberate simplifications (though, of course, all models are, to a greater or lesser extent). When building a model, mathematical biologists may start with the simplest possible scenario and gradually add complications as appropriate. Not all information about the variety of initiating conditions, constraints, and mechanistic bases of cancer is necessarily of relevance to a dynamical model of cancer progression. A model may be very effective at representing one aspect of a dynamic process or addressing one very specific question.

As Wimsatt (1987) has argued, "false models" may lead to "true theories." That is, deliberate simplifications of complex processes can yield important insights about what outcomes can be expected, given specific sets of conditions. It is permissible, and even sometimes advisable, to deliberately misrepresent the system of interest. Many models contain overt and explicit falsehoods. For instance, one can define equilibrium conditions (when it's known that in nature such conditions never or very rarely obtain), or one can identify what is expected for extreme cases on a continuum. Such models may provide a baseline or "null" case. Whether a model is "successful" ultimately depends upon what one wishes to use a model for. What matters to modelers, in other words, is not (or not simply) whether a model represents the system of interest in all respects, or with the greatest possible "realism," but whether the model can answer the question posed, provoke new lines of investigation, or suggest hypotheses worth exploring further. No model could possibly represent all possible cancers. The aim of mathematical modeling of cancer is often conditional generalizations. That is, under given conditions, what might we expect for systems of this type?

There are deterministic and stochastic models, individual-based or "cellular automaton" models, optimality models, models of competition, spatial dynamics, and hierarchical dynamics. One can use a model to explore different questions about different kinds of populations of cancer cells with different hierarchical structures. Such models are used to pose different kinds of questions, test various hypotheses, and simulate different aspects of cancer progression. Over the course of the latter half of the twentieth century, as new information about cancer heterogeneity, patterns of incidence, modes of initiation, and patterns of progression became available, new models addressing more sophisticated questions were developed.

The first mathematical models of cancer's dynamics were developed long before the molecular mechanisms or genetic mutations associated with cancer were well understood. Starting in the 1950s, these models represented cancer progression as a rate-limited multistage process, drawing upon epidemiological data—specifically, patterns of cancer incidence. For instance, Armitage and Doll (1954) developed dynamical models that predicted patterns of cancer incidence. They argued that since cancer incidence increases with age, cancer progression is a product of a multistage, rate-limited process of acquisition of mutations. Today, models of cancer progression are more sophisticated and involve the use of different kinds of mathematical tools—ranging from ordinary differential equations to agent-based approaches to elaborate simulations of spatial growth dynamics. Investigating such modeling practices can help us gain a better sense of the variety of functions theoretical models and model-building serve in the biomedical sciences, as well as the role of evolutionary thinking in cancer research. I begin by considering two examples from the recent literature: a simple model of competition between two cell populations and a more complex model of the evolution of chemotherapy resistance.

The former model is applicable to early stages of development of two "proto-cancer" cell populations. Most such populations do not progress to invasive disease. This model explores what features enable some cell populations to succeed. The second model examines how the evolution of chemotherapy resistance is dependent on a variety of factors: treatment with one or more chemotherapeutic drugs, population size, birth and death rates of cells, and mutation rates. In the following section, we will see how various models of cancer have been applied to specific observations or how data has been brought to bear on the models.

The simplest possible model of cancer competition dynamics represents two populations of incipient cancer cells; these populations might vary in a number of ways, but we'll consider the case of "stable" versus "unstable" populations. Stable cells have wild-type or relatively normal somatic mutation rates. Unstable cells have the "mutator" phenotype—meaning they are characterized by a lack of appropriate DNA repair mechanisms and so have elevated rates of mutation. They accumulate mutations faster than normal somatic cells. Under what circumstances do such unstable cells come to dominate stable cells? In order to answer this question, we need to know some of the fitness trade-offs associated with stability versus instability. We might predict that stable cells with intact repair systems face a cost; DNA repair takes time. Cell-cycle arrest and repair result in an overall slower rate of growth. On the other hand, unstable cells suffer from high levels of DNA damage—they bear a larger proportion of mutations, many of which would be expected to be deleterious. Using a set of differential growth equations, one can represent competitive interactions between such populations as a function of their intrinsic replication rate, mutation rate, and rate of DNA repair. Such models yield a variety of interesting results. As Wodarz and Komarova explain, "If the intrinsic replication rate of the mutator (M) is higher than that of the stable cells (S), then a high DNA hit rate can select for stable cells (S)," (106) but a low DNA hit rate selects for genetic instability (M). The reverse is the case if the intrinsic growth rate of stable cells is higher than that of unstable cells. In other words, we can use such models to represent the relative fitness (here measured simply as a matter of relative rates of replication and persistence of one type versus another) of different cell types, given the costs and benefits of high rates of mutation as compared with low rates.

Of course, this is a very simple model, and we also might wish to consider a variety of further complications, appropriate to different contexts or stages of cancer progression. In early stages of cancer progression—for example, before an incipient population of cells becomes a tumor—genetic instability may be relatively more advantageous than in later stages. This is because "mutators" might be more likely to acquire mutations that enable escape of the variety of controls on cell division in the tumor microenvironment—for example, apoptotic signals. On the other hand, too many mutations will result in a highly unstable cell, and perhaps eventually in high rates of cell death.

More sophisticated models can represent the emergence of drug resistance in a tumor or in leukemia for one or more drugs. Such models might start with a simple set of assumptions: for example, we can imagine that the population size of a tumor or leukemia cells in the body is N, the growth rate of cells is L, and the death rate is D. $L > D$ corresponds to clonal expansion. Mutations can lead to the generation of cell types that are resistant. Let the mutation rate be u; assume that resistant cells proliferate and that the rate of death of the wild type (non-resistant) is H. We can use this simple mathematical model to predict when and how quickly resistance will evolve in one specific type of leukemia (chronic myeloid leukemia, or CML), depending on the stage of growth or given treatment with one or more drugs. This can help us predict when treatment failure is likely, and so which kinds of combination therapy will be most effective and at what stages. Chronic myeloid leukemia has three stages: a chronic phase, (which is unsymptomatic), the accelerated phase, and the blast crisis. The latter two stages are associated with a very rapid rate of increase in undifferentiated cells.

On the basis of these models, Komarova and Wodarz (2005; Wodarz and Komarova, 2014) have shown that depending upon the initial population size of cells in a cancer, treatment failure is expected to occur when turnover rates of cells are high and mutation rates are high. Larger tumors evolve resistance more quickly, even with a relatively low rate of turnover. On the other hand, resistance can arise at lower tumor sizes if the rate of turnover of cells (D) is high. They predict that combination therapy will prevent treatment failure except when a cancer has a high turnover rate, or resistance mutations can be generated at rates several orders of magnitude higher than the physiological mutation rate. They extend the model to cases where there is complex tumor architecture (tumor stem cells), as well as cross resistance to multiple chemotherapies. In fact, the model was predictive: treatment schedules that maximize the chances of successful therapy are those that target different sizes of tumors or stages of CML. The model could be extended to consider further complications, such as tumor stem cells or hierarchical structure of cell lineages in a tumor.

What do these models illustrate about (a) the value of approaching cancer from an evolutionary perspective and (b) the nature of modeling in science?

There are several general conclusions we can draw from such cases. First, as already mentioned, cancer can be understood as a dynamic process of change in populations of cells undergoing complex interactions with their surrounding environment, involving changes in the proportions of distinct phenotypes and genotypes over time. At the same time, we know that this process is far more complex than the simple dynamic represented by the model. Second, the evolutionary perspective is, of course, a kind of idealization. Cancer is far from the "paradigmatic" case of adaptive evolution. Obviously, "successful" cancers eventually kill the patient (except in the case of some "infectious" forms of cancer, found in Tasmanian devils, for instance). Nonetheless, thinking about cancer from an evolutionary perspective can help us discover new hypotheses about the dynamics of cancer progression that are worth testing. The last case illustrates that modeling is a process of picking out what we take to be causally significant factors given some target or phenomenon of interest, whether chemotherapy resistance or the emergence of hypoxic cells. Of course, each cancer is genetically and epigenetically unique, and each cancer's environment is likewise unique. So while some of the conclusions of these models can be generalized across similar cases with shared features, the results of the models cannot be expected to generalize across all cancers.

What kind of knowledge do we gain from using evolutionary modeling to represent the dynamics of the emergence of these traits in cancer? First, at minimum, it is predictive knowledge. That is, we can use evolutionary models to predict and describe the dynamics of populations of cancer cells, as well as the emergence of complex tumors as in part a product of dynamic interactions between cells, cell lineages, and the tumor microenvironment and in part a byproduct of organismic adaptations. That is, these models help predict when and why cancers progress or fail to progress. Second, if one takes selection and drift to be causes of changes in populations over time, these models are not merely descriptive, but also explanatory. That is, to the extent that these models are accurate representations of cancer, they provide not just predictive but also causal knowledge.[5]

What kind of causal knowledge? Sober (2011) pointed out that in the actual practice of science, prior to whether we can know if X did cause Y, we often ask whether X tends to promote Y, or X would promote/cause Y. Sober (2011) calls answers to these questions "a priori causal models" and illustrates his argument with some examples from classical population genetic theory. For instance, all else being equal, if a heterozygote is superior to either of two homozygous

---

5. In my view, Stephens (2004), Millstein (2006), and Riesman and Forber (2005) (among others) have argued persuasively that selection and drift may be viewed as causes of evolution. Such a view is consistent with a variety of theories of causation. Addressing this debate at any length, however, would go well beyond the scope of this book.

alternatives, we should expect a stable genetic polymorphism to be maintained in a population. Such claims, Sober argues, are not trivially true as a matter of logical necessity, nor are they deductions from laws and initial conditions, as there are no empirical laws in the antecedent of the if–then conditionals that generate the "would promote" statement. Moreover, such "would promote" statements are true even if the antecedent never occurs. The function of such population genetic reasoning is to arrive at "how necessarily" theses, as well as "how possibly" hypotheses—that is, they show why some outcome must come about, provided a set of initial conditions is in place, and they suggest hypotheses to test. As Lewontin has similarly argued, "The delineation of the prohibited and the possible is the function of population genetic theory. The revelation of the actual is the task of population genetic experiments" (1985, 11).

Dynamical models of cancer often seek to answer questions about "the prohibited and the possible," given certain assumptions about the systems in question. The answers to such questions often have the form of "invariances" (cf. Woodward, 2003): both general and specific dependency relations between general properties of a population of cells, lineage, or whole tumor and outcome. For example, consider the following discussion of an optimality model for the evolution of genetic instability in cancer. At some stage in the progression of cancer, cancer cells acquire a variety of changes (microsatellite and chromosomal instability) that lead to elevated rates of mutation, errors in replication, loss of chromosomes, translocations, and so on. Genetic instability is a "two-sided" sword for the cancer phenotype: "too much" instability is detrimental to cancer cells and increases death rates; "too little" instability slows the progress of cancer. So Komarova et al., (2008) built a model to predict when instability is likely to be "optimal" for a cancer. They explain their modeling strategy as follows:

> By solving this problem, one can obtain valuable information about the growth of cancer. This is similar to the general philosophy of evolutionary game theory . . . The ideal (optimal) strategy may not even be realistic (there are many constraints in nature which escape modeling, but can make a strategy impossible). What occurs in reality, however, tends to approximate an ideal strategy. Finding the 'evolutionary stable strategy' or the 'Nash equilibrium' in the system helps us understand the general experimentally observed trend. A plausible explanation for the survival of those animals is that they have won the game against other animals that used an inferior strategy.
>
> In the present paper, we use similar ideas. We solve the optimization problem for cancerous growth and find optimal strategies. Does cancer always use an optimal strategy? Probably not . . . However, a cancer that follows the general trend, i.e., a strategy close to the optimal . . . will grow

faster. These are colonies that 'succeed' in causing a disease. Other precancerous colonies of cells might exist in any organ . . . but if they do not use a strategy sufficiently close to optimal, they do not succeed. (Komarova et al., 2008, 117)

Komarova et al. are asking a question about what "would" be optimal for any cancer. Their model shows why and how instability is beneficial at an early stage in progression and becomes a liability later on. The adaptive advantage falls off over time, because early on, genetic variation provides a pool of variation for adaptive evolution to draw upon. Later, this becomes a liability, for the simple reason that most mutations are disadvantageous, and at a certain point extreme chromosomal and genetic instability is a liability: cells cannot replicate at all if they bear so many mutations that they cannot perform meiosis. If we see what we would expect or what is predicted by a model, and the initial conditions of the system (roughly) map onto that of the model, then we have good reason to think that the model is accurate in capturing the essential causal factors at work in this system. But of course no system is going to be like the model in all respects.

By way of another example of a highly abstract model, Spencer et al. (2006) built a three-dimensional, agent-based computer simulation model that represents cancer progression as a stochastic process. Each cell is treated as an autonomous agent that follows simple rules governing its behavior, and behavior changes as it acquires mutations. The simulation explores questions about the timing and order of onset of different mutation types in the progression of cancer. It demonstrated that early-onset cancers proceed through a different series of mutation acquisitions than late-onset cancers. Specifically, the acquisition of genetic instability early in cancer progression makes for more heterogeneous tumors, and thus more early-onset ones. Cancers that first acquire limitless replicative potential (e.g., via lengthening of telomeres) have a more delayed onset. Heterogeneity varies with the timing of acquisition of genetic instability (as might be expected). In other words, these computer simulations enabled researchers to identify certain invariance relations between initial conditions and outcomes.

These modeling strategies are not unlike Sober's characterization of "would promote" statements in population biology. Sober claims that such "would promote" statements are not trivially true; nor are they merely logically necessary. Rather, he claims that they are both causal and necessary. This may strike one as initially implausible: How can a causal claim be "necessary"? Hume long ago argued that causal claims are empirical, and thus cannot attain the status of "necessity"—we observe no "necessary" connections in nature, only empirical regularities. These "would promote" statements do describe empirical facts and empirical regularities, but they are not known to be true by direct observation. Rather, they are consequences of mathematical reasoning about relationships between values of variables, themselves abstracted from the phenomena. Such variables are often "generic" (population size, fitness, mutation rate) rather than specific, and permit one to derive general conclusions about the behavior of any evolving population, that is, any that meet the conditions set out in the models.

Many may find calling such if-then generalizations "a priori" rather counter-intuitive. But Sober does not mean by "a priori" to suggest that such propositions are somehow known prior to experience. Rather, they are known via demon-stration (rather than observation) and have the status of "necessity," given that they involve mathematical derivations of conclusions from initial conditions. We come to know them by logical or mathematical inferences, and they are "nec-essary" consequences of the assumptions we use to construct our mathematical models. They allow us to predict what we "ought" to expect of a system, if and when the system maps onto the model.

Several philosophers have taken issue with Sober's argument. For instance, Lange and Rosenberg argue that Sober's "would promote" statements are *neces-sary but not "causal."* Their argument is that either statements such as Sober's are "shorthand" for particular facts, or they are "conditional[s] in which the ante-cedent turns out essentially to involve a mathematical expectation and the con-sequent turns out to be the very same expectation."(2011, 596) If the former, then the statement is made true by empirical investigation; if the latter, then the statement is true by definition, and so is not causal. Lange and Rosenberg (2011) argue that we need to know the actual causal differences that yielded the out-come in a given case to explain an outcome. Thus, in their view, Sober's "would promote" statements are neither causal nor explanatory. But notice how they frame the problem: we explain *particular* outcomes and *particular* facts, drawing upon such shorthand descriptions. Yet one way of understanding Sober's examples is that they are not concerned with particular facts or outcomes. Rather, they describe general invariances or general relationships that would ob-tain in any population sharing the same initial conditions (all else equal). Of course, in the biological world, not everything is equal. But what these models can do is describe general patterns or regularities and how they are governed by combinations of causal variables. Such generalizations tell us what to expect in any evolving system. Birch (2017) refers to Hamilton's rule, for instance, as an "organizing framework" for social evolution. Similarly, the kinds of invar-iant relationships Sober is describing are explanatory in that they identify high level generalizations about the fundamental regularities governing all evolving populations.

Consider the following example. Tomasetti and Vogelstein (2015) predict that tissues with higher rates of stem cell renewal are by and large more vulner-able to cancer than tissues with lower rates. That is, as a matter of chance alone, populations with more stem cell renewal will generate higher rates of mutation and so acquire more cancers. If we assume that mutation rates are constant in a given population, a larger population of dividing cells will acquire mutations more quickly than a smaller one. Is this claim made true in virtue of "causal" or "statistical" facts? In favor of causal, you might appeal to the following counterfactual: "If mutation rates are constant, and cells divide more often than not, they will acquire more genetic variation." And the general regularity that supports this counterfactual is "Rate of cell division, provided mutation rates are constant, predicts the extent of genetic heterogeneity in a population over time." This generalization may be represented mathematically as a necessary truth, since any rate-limited error-generating process will generate more errors, the higher the rate. The point is, we can generate a mathematical model that describes the regularity we find as a necessary consequence of a set of initial conditions. Tomasetti et al. are offering a kind of "theoretical" explanation of a general pattern, not a causal explanation of a particular event. Any populations of cells that meet the conditions described by Tomasetti et al. will display this regularity; higher rates of cell turnover yield higher mutation rates. By analogy, typing errors might be more common if one types quickly. That is, this might be viewed as a kind of "a priori," yet causal truth.

Lange and Rosenberg (2011) deny that there are such truths. But this denial is founded on a contested assumption. Lange and Rosenberg assume we must choose: either a truth is causal or it's a priori (necessary). It cannot be both. Sober's argument was aimed at contesting this very assumption; his examples were supposed to establish that the choice is a false one. Both mathematical analysis, and actual (particular) causal facts are relevant to and (arguably) equally necessary for explaining why cancer is in fact more common in tissue with higher rates of stem cell division.

In his piece on real statistical (RS) explanations, Lange (2013) insists that explanations in population biology that appeal to claims about drift are *non-causal statistical explanations*; they are made true in virtue of the fact that certain processes in nature are simply "chancy." An "RS explanation," he explains, "does not proceed from the particular changes of various results—or even from the fact that some result's chance is high (or low)" (173). Rather, such explanations appeal to the fact that some processes are subject to chance, and so "by chance alone" we get a run of heads or tails, or an extreme outlying case, or a regression to the mean. Such outcomes may be an instance of a "signature statistical process," such as "regression to the mean," or may instantiate a "random walk." Lange takes fluctuations in gene frequency due to sampling error from one generation to the next (drift) as an exemplary case.

Such statistical explanations that appeal to drift depend upon or instantiate, in my view, causal "would promote" statements very similar to what Sober (2011) has in mind. The isolation of small subpopulations is, *in general*, a cause of reduced heterozygosity in a population. Such population-level claims meet

Woodward's (2003) minimal conditions on causation: a change in the value of one variable would systematically change the value of the outcome variable. But, of course, they are modal claims about general population level variables rather than empirical claims about actual populations. This fact about heterozygosity in evolving populations in some sense depends upon a statistical generalization that smaller sample sizes have less variability, but it is not reducible to the statistical fact. When we appeal to isolation as a "cause" of reduced variation, we are appealing to changes in real population variables that change certain outcome variables. That is, though this claim can be represented as a statistical "a priori" truth, it is also a causal claim; the small sample size does cause, and also explain, the reduced heterozygosity.

Sober's statements have the appearance of necessity because they are so stable, or are invariant generalizations about any evolving system. When they describe the relationship between population-level variables like population size and heterozygosity, where change in one value predictably changes the other, they are causal. Likewise, in my view, even though Lange's examples are (in part) made true by facts of probability theory and statistics, the conditions that make them true are causal conditions. Isolating a small population is thus a cause, in this sense, of reduced variation since it has a relatively stable tendency to promote reduction in genetic variance.

Dynamical models of cancer similarly make certain assumptions about how best to mathematically represent the systems in question and derive "would promote" statements of the sort described by Sober from various initial conditions set out in the model. What makes such "would promote" statements causal is that the models accurately represent the causal variables that make a difference to various outcomes in cancer progression. What makes them "necessary" is that given the mathematical representation, conclusions follow from initial conditions as a matter of mathematics. They are, in other words, both causal and "necessary." Just as in the Tomasetti case described earlier, both facts are relevant to the explanation of the facts to be explained. That these models depend upon mathematics or the probability calculus for their necessity does not make them less empirically relevant, for without the probability calculus, we would be missing a central factor relevant to explaining the dynamics of cancer progression.

The answers to "what if" questions about neoplastic progression often have the form of "invariances": both general and specific dependency relations between properties of a population of cells, lineage, or whole tumor and outcome. Models such as these both predict and explain various outcomes by appeal to invariance relationships, generalizations about causal or dependency relations between initial conditions in populations of cells and expected outcomes. Two examples are the following: "If all cancerous cells were susceptible to the drug, then therapy would inevitably lead to the eradication of the cancer"; and "The higher the value of u (the rate at which resistant mutants are generated), the lower the tumor size at which treatment fails." Such invariance relationships have a dual status: they are derived mathematically or probabilistically from the setup of the model, but they are not simply this: they also represent causal relationships between properties of populations. That is, mutation rates or population sizes are not simply background conditions but causes of evolutionary outcomes. If, for instance, one varied population size independently of anything else, then the rate of cancer progression would change; larger populations would more reliably (on average) yield more variation for selection to act upon, and thus for novel adaptive mutations to arise and spread. Thus, to the extent that these models are analogous to the conditions on causal intervention that Woodward, Pearl, and others have described, where we imagine we might intervene on one variable or another independent of others, they are causal and explanatory. Of course, in order to generate such models, we need to know which causal variables are, in general, significant.

Which causal factors are significant? If anything has become evident over the past several chapters, it is that the answer to this question depends upon the grain of analysis and empirical details about the system of interest. The relevant contrast to the target in such explanatory questions determines the relevant grain of analysis. Zooming out, we can make general claims about causal invariances that hold of any cancer. Zooming in, targets of explanation may be more precise and specific to a subset of cases with distinctive features. Depending on the grain of analysis, or the temporal and spatial scale we are exploring, different details about causal subsystems may be more or less explanatorily relevant to one's target explanandum. When building a model, biologists often focus on the big picture or choose to designate a class of systems of a type. Alternatively, biologists might focus on one subsystem or one set of causal relationships or invariance relations. Modelers' decisions about what to include in the model are in part shaped by empirical understanding of the system in question and in part by the questions they ask about the system or the kinds of outcomes they are interested in investigating.

Models such as those discussed in this chapter help scientists discover invariance relationships, generalizations about causal or dependence relations between initial conditions in populations of cells, and different outcomes. Such invariance relationships serve a variety of functions: analogous to laws, they make predictions, they frame future research, they explain what we do find, and they function as theoretical frameworks, shaping background assumptions about expectations for what "should" or "would" be likely to occur. In other words, they constrain plausible explanations of what in fact does occur in cancer progression. All these outcomes are the product of building models and studying how these models behave over time, given different choices of parameters and/or values of variables. Building a model, however, involves a series of choices about what variables are exogenous and what endogenous, what is held fixed, what is allowed to vary, and what scale, scope, or level of detail to include. In other words, the relationship between model and world is often (at best) an approximate and hypothetical one. We can often achieve the same dynamic with a different model.

Empirical tethering or "fit" to the world is not an "either/or" but rather a "more or less" matter: models must be empirically tethered to known causal outcomes or known inputs, but the role(s) of subsystems, feedback, and other intermediate effects are often black-boxed. Sometimes such black boxing is unproblematic: Some general dynamical models are silent about the subsystems in the model. As long as those subsystems serve the same function in the larger model, they can be "black-boxed." However, when and how (much) one can permissibly black-box in a model will depend upon the question, and so upon the degrees of freedom one has in choices involved in building a model.

6 EXPLAINING CANCER

6.1. Introduction

In the third chapter of his *Structure of Scientific Revolutions*, Kuhn remarks:

> It is no criterion of goodness in a puzzle that its outcome be intrinsically interesting or important. On the contrary, the really pressing problems, e.g., a cure for cancer or the design of a lasting peace, are often not puzzles at all, largely because they may not have any solution. Consider the jigsaw puzzle whose pieces are selected at random from each of two different puzzle boxes. Since that problem is likely to defy (though it may not) even the most ingenious of men, it cannot serve as a test of skill in solution. In any usual sense it is not a puzzle at all. Though intrinsic value is no criterion for a puzzle, the assured existence of a solution is. (1962, 36–37)

This passage contains an important insight, both about science and about cancer research in particular. A central criterion of a good puzzle is that there is a solution. Curing cancer and designing lasting peace are both intrinsically valuable projects. But they are not single puzzles; they are each *sets* of *distinct* puzzles. There may never be a cure for cancer; there may, however, be cures. Likewise, there may never be an explanation for cancer; there are, however, explanations.

When I first began this book, my intention was to explore how cancer research sheds light on theories of explanation in the philosophical literature. Thus, this final chapter initially consisted of an overview of various philosophical theories of explanation, followed by a discussion of how explanations of cancer served as cases of one or the other model of scientific explanation. The closer I looked at the practice of cancer research, however, the less inclined I was to see explanation as the most pressing goal of cancer research; instead, it seemed that the vast majority of cancer research is directed at prediction and control.[1]

Moreover, what explanations there were, were by large fine-grained. In other words, the vast majority of cancer research is better described as addressing *particular* questions about patterns, processes, and aspects of cancer, rather than arriving at "general theories" of carcinogenesis. This conclusion runs contrary to one very intuitive picture of science. On this view, the aim of scientific inquiry is arriving at true general theories, and successful explanation is guaranteed, provided we have the right theories. Explanations on this view are "theory driven": theories are intended to subsume or "cover" the facts to be explained. This view of science is often aligned with a particular view of scientific progress and rationality: science is a cumulative process of achieving an ever more systematic picture of the natural world, one that properly classifies all the kinds of things in nature, their properties and lawlike behavior. Natural kinds all have some common fundamental features—whether common microstructure or common causal basis. It is these shared properties of kinds that allow us to make general predictions and give explanations in light of our theories.

If cancer were a natural kind such as this, then we might expect scientific research on cancer to work like this. But as argued in Chapter 1, the more progress we make in cancer research, the less plausible this picture becomes. Cancer is not a single disease, but a massively heterogeneous class of disease processes. There are associations between various properties of cancer types and subtypes and various outcomes, and we can classify and cross-classify cancers in light of these associations. While there are some features broadly shared by many cancers, there are many respects in which cancers are very different. And, there are few if any lawlike generalizations about cancer.

So, despite rhetorical appeals to the "war against cancer," cancer is not a single fact to be explained or a single problem with a single solution. As in much of biology, the phenomena can be decomposed in different ways, leading to different questions.[2] In such circumstances, scientists need to make strategic choices about which questions they are in the best position to answer. How exactly do scientists decide which puzzle is deserving of their attention?

Over the past twenty-five years or so, historians and sociologists of science have argued that advances in technology and in the pace of publication, as well as the growth of "big" science and "big data," have transformed the culture of

---

1. And, in any case (as a kind reviewer for OUP pointed out), there was already an extensive body of philosophical literature on the history of theories of scientific explanation (see, e.g., Woodward, 2003, 2014; for a discussion of explanation as it pertains to biomedicine see, e.g., Braillard and Malaterre, 2015).

2. See, e.g., Wimsatt, 1972; Bechtel and Richardson, 1993.

science, leading to an emphasis on "solvable" problems (i.e., problems a postdoc can solve and get a publication out of), and "getting the right tool for the job" (i.e., easily transferrable technologies and methods, from PCR (polymerase chain reaction) to model organisms). Philosophers of science have paid less attention to this history and sociology of science than they ought to, in my view. If they did, they might conclude that while sometimes choices of problem are dictated by interest alone, more often they are dictated by what we might call "expedience." That is, which question a scientist is likely to ask (as well as how they go about addressing that question) is informed by their disciplinary training, associated research traditions (methods, laboratory techniques available, etc.), institutional support, and opportunities for funding or publication. Hope for downstream impact may certainly inform such choices, but often, it is the tools available that determine which puzzles scientists choose. Often it is simply that we have the right tool available for the job (Keating and Cambrosio, 2003, 2011; Fujimura, 1987, 1988, 1996).

Over the course of the twentieth century, different cancer scientists coming from different disciplinary perspectives have focused their attention on different puzzles. Their choice of puzzle often simply depended upon available tools. Cancer research was shaped by advances in a wide array of biological disciplines, incorporating new empirical data and methodological and technological innovations along the way (Morange, 2007, 2011; Cairns, 1976). Sometimes lawlike generalizations were proposed as explanations of patterns or processes observed across a variety of cancers. But cancer research very rarely looks like the picture of science according to which the aim is arriving at general theories.

What then does success in cancer research look like? I believe that Kuhn perhaps inadvertently suggests a very interesting answer, though he did not put it exactly this way: the aim is solving a puzzle.[3] Of course, this claim is terribly vague, but bear with me. The analogy is illuminating in several respects. First, note that the rules for what counts as a solution will vary with the puzzle. Doing a puzzle means committing to following the rules appropriate to that puzzle. A solution to one puzzle does not ordinarily count as a solution to another puzzle, though

---

3. I am not by any means committed to Kuhn's larger view about the nature of scientific change and, in particular, his view of "paradigm shifts." I believe Kuhn was still in the grip of a particular view of science inherited from logical empiricism. On this view, theories are far more central to scientific inquiry than practice—or experiment and intervention. In my view, both are essential to science and scientific change. On Kuhn's view, scientific change involves shifts between competing worldviews, metaphysical commitments, and associated theories. I don't accept that view of scientific change; change in science is often far more piecemeal and gradual. Kuhn thought that one theory or paradigm must "win" and the others lose. One of my arguments here is that this view of science is misguided.

sometimes puzzles and their solutions may be mutually informative. Puzzles are discrete, and expected to have discrete solutions. We typically do not criticize someone for failing to solve one puzzle when he or she sets out to solve another. Committing to a puzzle as interesting does not require denying that there are other puzzles worth investigating. In this chapter, I explore two central puzzles that have driven cancer research in the latter half of the twentieth century: Why does cancer incidence increase as we age? And, why don't we get cancer more often than we do?

6.2. Puzzle 1: Why Does Cancer Incidence Increase as We Age? The Rise of the Multistage Theory and the Path to Oncogenes

The path to what Fujimura (1997) called the oncogene "paradigm" has been far from straightforward. However, one observation arguably put many researchers on the path to what became the "multistage" theory, and eventually, the search for "oncogenes": cancer incidence (for the most part) increases with age. Why? There are several ways of going about answering this question, but the way available to scientists when very little empirical data are available is mathematical modeling. So, what cancer researchers did in the 1940s and 50s was ask: What would a mathematical model look like that would account for this pattern?

Starting in the 1940s, biologists began to develop a family of mathematical models that represented cancer initiation and progression as a product of a series of rate-limited steps. Charles and Luce-Clausen (1942) noted that mice acquired skin tumors after repeated application of benzopyrene, and so proposed one of the first quantitative models of cancer as the product of multistage progression. Skin tumors on mice painted with benzopyrene arose only after multiple applications, suggesting that there might be a series of changes the mice needed to pass through to produce a cancer. A variety of scientists, with several disciplinary backgrounds, elaborated upon these models of cancer as a multistage process. They drew upon a variety of evidence relevant to explaining average age of incidence curves including but not limited to studies of the effects of chemical carcinogens on animals, and patterns of inheritance of childhood cancers.

While the early models were simple, idealized representations of cancer initiation and progression, intended primarily as hypothetical, they became more complex, integrating a wider array of data as it became available. Over the course of the twentieth century, it became clear that a variety of remote causal factors (e.g., environmental carcinogens such as smoking) and proximate ones (e.g., chromosomal abnormalities typical of cancer cells) were of relevance to patterns in average lifetime cancer incidence. However, it is of utmost importance to be clear that, at least initially, these models were concerned to account for a pattern in cancer incidence, not posit mutually exclusive hypotheses about carcinogenesis. The target of these multistage models of cancer was average lifetime *patterns of cancer incidence*. The modelers did not intend their models to explain *everything of relevance* to carcinogenesis. In evaluating the success of these solutions, then, we need to be attentive to the target.

Doll, one of the early modelers, along with Sir Bradford Hill, was one of the first epidemiologists in the 1950s to note that on average, lung cancer incidence gradually increases with age, peaking at about 55-65, and eventually leveling off through old age. They noted that smoking appeared to shift the lifetime risk to earlier ages. Armitage and Doll (1954) argued that a series of multiple "hits" or mutations, caused by irritation of the lungs in smokers, could yield shifts in curves of cancer incidence to earlier ages, especially with longer duration and more extensive use of tobacco. Their model was as follows: cancer incidence increased roughly by t^{n-1}, where t is age and n is the number of rate-limiting steps. Fitting the data to the curve of acceleration of cancer incidence (a derivative of the incidence curve) for a variety of cancers, it appeared that at minimum, $n \approx 6-7$ events. Nordling (1953) and Stocks (1953), who generated log-log plots of incidence for different cancers, used these models of differential incidence to argue that *different cancers required different numbers of steps for onset*.

While different cancers have slightly different patterns of lifetime incidence, what these modelers took to be most striking about these age-specific incidence curves was similarities in the general shape of the curve. That is, whether a cancer is common or rare, in the breast or the lung, average age-specific incidence curves have a similar shape; cancer incidence increases by and large as a function of age. That cancer incidence by and large increases as a function of age *suggested* a stepwise, cumulative process underpinning this outcome. One of the core insights driving the development of the multistage theory is that studying *patterns* of cancer incidence at the population level can potentially yield insight into the *processes* of carcinogenesis.

Starting in the nineteenth century, chemical agents like dyes, soot, coal dust and tar, and, eventually, X-rays and radium were found to induce cancer in experimental animals and in exposed workers, such as chimney sweeps, mine workers, and factory workers, such as the women who painted watch hands with radium dyes. The mechanism by which these agents caused cancer, however, was still unknown. The induction of cancer in experimental animals via mutagens like tobacco tar and radium, was, however, taken to be suggestive of a link between mutagenesis and carcinogenesis. That is, many supposed that the underlying rate-limited process was the accumulation of mutations to cells. The idea that cancer could be a product of mutations can be traced back to the very beginning of the twentieth century: it was first prompted by observations of chromosomal abnormalities in tumor cells (Boveri, 1914, 1929).

This model of cancer as due to a specific number of "hits" was then extrapolated to a novel context: patterns of familial inheritance of cancer. As discussed in Chapter 1, retinoblastoma is a relatively rare form of cancer of the retina, which appears either unilaterally (in one eye) or bilaterally, typically in children, and appears to be strongly inherited. In 1971, Knudson generated a very precise hypothesis, sometimes called the "two-hit" theory: "that retinoblastoma is a cancer caused by two mutational events. In the dominantly inherited form, one mutation is inherited via the germinal cells and the second occurs in the somatic cells. In the nonhereditary form, both mutations occur in somatic cells" (820). Drawing upon patterns of familial incidence of retinoblastoma (siblings with and without the disease) and marking the patterns of age of onset of bilateral versus unilateral cancers, Knudson developed a mathematical model that represented cancer as a product of at least two mutations to a dominantly inherited gene. Knudson's prediction was borne out. In fact, *RB* (the retinoblastoma gene) was the first and prototypic tumor "suppressor" gene, a gene associated with regulation of the cell cycle and apoptosis. Knudson developed this hypothesis on the basis of patterns of age of onset; the patterns of incidence constrained the available hypotheses. By comparing children who acquired the disease earlier versus later, and in two eyes versus one, he hypothesized that the earlier-onset and more devastating cases were likely due to the fact that these children needed to pass through fewer steps to get cancer. Knudson's *RB* gene was later identified, and its mechanism of action and role in the onset of retinoblastoma are well understood (Kumar, 2010).

The multistage picture of carcinogenesis pointed to a common causal basis for both familial and sporadic forms of the disease, as well as familial patterns of inheritance and age of incidence curves for sporadic or non-familial cancers. In other words, data from very different sources was brought to bear on the construction (and reconstruction) of a family of mathematical models that proposed a common, rate-limited causal process as underlying all forms of cancer. Each source of data supported the core idea behind the models, that cancer was the product of a process extended over time, limited by the accumulation of stepwise changes (namely, mutations). The multistage model drew upon data from a variety of sources: cell biology (identification of chromosomal abnormalities in tumors, suggesting that mutations may be involved in cancer), toxicology (tests on animal models suggesting a "multiple-hit" theory), epidemiology (evidence that the curve of incidence could be shifted to earlier onset in smokers, suggesting that the rate could somehow be increased), and eventually, clinical medicine and studies of hereditary incidence of cancer (yielding Knudson's hypothesis and the subsequent discovery of the *RB* gene).

Each of these independent sources of data informed the construction of the models of cancer's dynamics.

Of course, modeling often involves idealization and simplification; for instance, one might assume (falsely) that the rate of some process is constant or that a series of events occurs continuously and that each event is independent of the others. It is easier to build a mathematical model when one makes such simplifying assumptions, and one often builds upon a prototype, or a model of the same or a similar process with similar dynamics. The multistage model deployed both simplification and appropriation. For instance, some of the first models made a number of assumptions:

- Mutations are the exclusive rate-limiting events in cancer progression.
- Cell mutations are a Poisson process, or stochastic process, where each event is assumed to be independent of others and to occur in a given time interval, and a tumor might occur after k such events (Nordling, 1953).
- Cancer initiation is a chance event, takes place suddenly, and has a specified transition probability density function, or rate of change per unit time for each tissue type (Armitage and Doll, 1954).
- Each tissue type has a specified induction period, constant for all initiation events in that tissue but varying between tissues according to some distribution.

Some of these assumptions were known or suspected to be false; others were merely hypotheses at the time they were proposed. Armitage and Doll (1954) were quite candid, for instance, in granting that different individuals' responses to the same environmental insult might have different outcomes or that the rates of change in the same tissue type might vary by individual exposed.

By representing cancer's onset and progression as a rate-limited, multistage process, Armitage and Doll, among others, had proposed a solution to the puzzle as to why cancer incidence increases with age, and how and why there are different curves for different cancers. If cancer involves many steps, and these steps have a constant rate, we can understand cancer as the endpoint of the gradual accumulation of changes to cells over time. Variations between different curves of incidence for different cancers, or, different ages of onset, might then be explained by inherited mutations, or different environmental exposures, or rates of turnover of cells in different tissues and different tissue architecture, or all of the above. The "target" of explanation or "explanandum," of the multistage models is a general pattern, not the detailed causal mechanisms involved in setting the pace of these dynamics.

What role do these models play in explaining cancer? Arguably, the multistage models provide one of the closest things in cancer science to a deductive nomological model of explanation. But like any model built to fit the data, it has a somewhat tautological flavor. Let u equal the mutation rate, t equal age, and n the number of "hits" or mutation events necessary for a cell or cell population to yield cancer. It follows that the approximate rate (incidence) of occurrence at time t is proportional to $u^n t^{n-1}$. Individuals carrying a mutation develop inherited cancer after $n - 1$ steps. And the incidence ratio of sporadic to inherited cancers at any age t is $ut/n - 1$ (cf. Frank, 2007). The reason cancer incidence increases (by and large) as a power of age is that cancer incidence is a product of sporadic somatic mutations as cells divide over the course of our lifetimes. Inherited forms of the disease by and large arrive at earlier stages, because a "step" in the process is "skipped."

Of course, this is just a model; exceptions to the rule, much as with Mendel's laws, abound. For instance, some cancers have accelerated onset; these early-onset cancers due to mutations that cause high levels of chromosomal instability are sometimes characterized as due to "chromothripsis" (Stephens et al., 2011; Rode et al., 2016). Moreover, as we now know, mutations are certainly not the exclusive rate-limited process of relevance to cancer onset. Nonetheless, the multistage models were fruitful, in that they both solved one puzzle, and pointed the way toward more puzzles. Namely: What were the causes of rate-limited changes to cells? Were the causes of cancer cell transformation "intrinsic" to the cell, or "extrinsic," or both?

For some time in the 1960s and 1970s, a central contender solution to this puzzle was viruses. In 1909, Peyton Rous demonstrated that you could infect a chicken with a virus that led to cancer. Rous's work suggested that a tumor was induced by an infectious agent. Rous removed a tumor from the breast muscle of a chicken, ground it down with sand, filtrated it through a fine-pore filter, and then injected this filtrate into a young chicken, leading to tumors. This suggested that a very small infectious agent was causing the disease. Further evidence for this hypothesis came from the fact that in addition to being transmissible, the infectious agent seemed to be able to multiply in the host. Researchers were able to harvest more of this virus from the infected chicken than had been originally injected. The virus was later called "Rous sarcoma virus," or RSV.

However, the story turned out to be rather more complicated. In the 1960s, a postdoctoral student at Cal Tech, Harry Rubin, found that RSV introduced into cultures of chicken embryos reproduced indefinitely, displaying many of the features of cancer cells. Temin, another student at Cal Tech, showed that descendants of an RSV-infected cell continued to harbor genes (from the viral infection) that led to cancer. Howard Temin's work on RSV was especially radical, in that it challenged the then "central dogma" that information only flows "forward," from DNA to RNA. Temin was one of the first to suggest that information could flow backward, from the RSV into the DNA of infected hosts.

Retroviruses inserted into a host cell might make double-stranded DNA copies of themselves that were then integrated into the host's DNA and replicated when the cell itself replicated. Temin's proposal was initially met with a great deal of skepticism, but it eventually led to the technology used in almost every biology lab today: PCR (polymerase chain reaction). In addition, it's now considered commonplace that retroviral infection can lead to many kinds of cancers.

The idea that cancer induction could be studied in culture and that viral infection could be a cause of cancer led to the discovery of several classes of tumor viruses through the 1960s and 1970s, including human papillomavirus (HPV), which is responsible for most cervical cancers. However, it was still not well understood at this point why or how viral infections led to cancer.[4]

In the 1980s, using some of the first DNA probes (made possible by Temin's discovery of reverse transcription), Bishop and Varmus found evidence suggesting that the specific gene associated with viral cancer in chickens was not "viral" but was very common, shared not only by other vertebrates, like chickens, but also by arthropods and even sponges. That is, in its non-mutated form, the *src* gene was highly conserved (found throughout the evolutionary tree), and *c-src* thus came

---

4. Today infectious agents are believed to cause one-fifth of human cancers—HPV-induced cervical cancer, liver cancers induced by hepatitis B virus, and Burkitt lymphoma induced by Epstein-Barr virus (Weinberg, 2013). However, there is no uniform mechanism by which virally induced cancers come about. For instance, human T-cell leukemia virus (HTLV) causes cancer; however, not all infections with HTLV-I lead to cancer. Only 3–4% of those with lifelong infection by this virus develop adult T-cell leukemia. This is because the process is indirect; the virus produces a protein that activates growth factors, stimulating the proliferation of hematopoietic (blood) cells, which can in some cases lead to leukemia. Leukemia, in essence, entails an "overproduction" of white blood cells, and so by inducing proliferation in such cells, the virus can (indirectly) induce leukemia. In other words, the chain of causation in this case is indirect: but for the HTLV-I infection, this particular kind of leukemia would not have resulted. So there is a meaningful sense in which these cancers are "viral" cancers; but for the infectious agent, there would be no cancer. However, it is also true that but for the evolution of the virus to express the protein that activates growth-stimulating proteins and their promotion of hematopoietic cell proliferation, there would also be no such cancer. Viruses are *a* cause, but they are not the only cause of relevance to explaining the disease. Evolutionary history, particularly the coevolution of viruses and their hosts, plays an important role in explaining these cancers.

Examples such as this abound in cancer causation. Consider the other viral cancers; there are two main classes of viruses—DNA viruses, which carry DNA genomes, and RSV viruses, which carry RNA molecules. Each kind of virus produces a different kind of cancer. For instance, HPV is a DNA virus that leads to warts or papillomas on the skin and, occasionally, cancer. HPV is a member of a family of papovaviruses, which are able to integrate copies of their genomes into host cell DNA. Fully 99% of cervical carcinomas carry fragments of HPV integrated into their DNA. Cells transformed by these and other papovaviruses exhibit characteristics of cancer; they are "immortalized," or proliferate indefinitely in culture, and are anchorage-independent (their growth is not inhibited by neighboring cells). Thus, in the case of DNA viruses, but for infection, or perhaps but for the evolution of the capacity of DNA viruses to integrate themselves into hosts, there would be no such cancer.

to be called a "proto-oncogene." What explained its presence in RSV was that infection by this virus in a distant metazoan ancestor led to the incorporation of the proto-oncogene into the viral genome. Over time, it had been transformed or mutated into a "cancer gene." These transformed genes (originating in a vertebrate genome) can then induce cancer: infection by retroviruses leads to cancer via "insertional mutagenesis": when the virus infects the host, reverse transcription incorporates that same gene into the host's cells. The *src* gene then leads to cell proliferation and, ultimately, cancer. But the normal gene can *also* be subject to somatic mutations and thereby contribute to cancer. Varmus and Bishop won the Nobel Prize in 1989 for demonstrating that the viruses are simply the vector for a specific mutation in a highly conserved gene (a gene shared by many species in the tree of life).

In other words, perhaps ironically, the first "cancer gene" was discovered as a result of the investigation of viruses and their role in cancers. Moreover, the discovery was made possible by challenges to the central dogma. Throughout the 1980s and 1990s, molecular biologists raced to discover more "oncogenes" and "tumor suppressor" genes, and occasionally promoted their research by suggesting that once we found these "tumor genes" we could "turn cancer on or off" at will. Of course, so-called cancer genes are not genes "for" cancer so much as genes that play a role in complex pathways that either promote transcription or "advance" the cell cycle, or "halt" transcription and cause cell death, among many other roles. In other words, there are no genes for cancer so much as there are genes involved in the healthy or otherwise normal functioning of the cell cycle, which either are deleted or epigenetically altered (e.g., hypomethylated or hypermethylated) in ways that push the "accelerator to the floor" of the cell cycle or serve as "brakes on" mechanisms that would otherwise lead to apoptosis.

In sum, the story of the multistage theory is also the story of the rise of the oncogene paradigm. It was the success of the multistage theory that led to the idea that a common rate-limited process underlay all cancers. This process was, eventually, taken to be mutations that occur during somatic cell division. The somatic mutation theory is the view that cancer is essentially the product of the gradual accumulation of cell-intrinsic changes: mutations, during somatic cell division. Of course, as discussed in Chapters 1 and 3, we now know that cancer is not due exclusively by cell-intrinsic causes, but many different factors in interaction. So while the oncogene picture has been useful, it is focused exclusively on one very local and specific temporal and spatial scale. Nonetheless, the initial puzzle was fruitful, yielding a family of puzzles: How did these inherited genes play a role in cancer? What role exactly? Are some mutations more common than others? Do they occur more frequently in some cancers than others? The oncogene paradigm, however, also yielded a number of paradoxes: in particular, given that cell division is ubiquitous, and mutations are likely to accumulate over time in cells and tissues throughout the body, why don't we get cancer more often than we do?

6.3. Puzzle 2: Why Don't We Get Cancer More Often?

According to the multistage theory, cancer is a disease of disorderly cell growth, caused primarily by mutations to cancer cells. These mutations specify the aberrant phenotypes of cancer cells. These "hallmarks" of cancer are all taken to be due (primarily) to changes in gene expression: "[C]ancer [is] a disease involving dynamic changes in the genome. The foundation has been set in the discovery of mutations that produce oncogenes with dominant gain of function and tumor suppressor genes with recessive loss of function" (Hanahan and Weinberg, 2001, 57). But cells are constantly dividing, and mutation is ubiquitous. Why don't we get cancer more often than we do?

The question itself can be broken down into yet further questions:

- What factors prevent the advancement of cancer?
- What role, if any, does the tissue microenvironment play in cancer? How, if at all, does the extracellular matrix and/or the cytoskeleton either prevent or promote cancer? How does tissue architecture prevent cancer?
- What role (or roles) does the immune system play in preventing cancer?
- How does cell metabolism play a role in halting or advancing cancer?
- How does stem cell architecture play a role in halting cancer?
- What evolutionary explanations can we give for the relative infrequency of cancer, given the frequency of mutations as cells divide?

These questions were among those (of many) that led Warburg (1956), Bissell et al. (1982), and Soto and Sonnenschein (1999) into research on cancer. The investigation of these puzzles drew upon the resources of a variety of disciplinary perspectives, over the course of the late twentieth century: developmental biology, physiology, genetics, immunology, and evolutionary biology.

Warburg was a biochemist, physiologist, and physician, who noted something striking about the metabolic features of cancer cells: they produce energy by a relatively high rate of glycolysis (almost two hundred times higher than their cell of origin) and then engage in lactic acid fermentation. That is, cancer cells have an unusual anaerobic (or oxygen-free) metabolic process, compared with the aerobic process typical of normal cells. Warburg proposed a hypothesis to explain this and, indeed, to explain the origins of cancer, generally. It is worth quoting his classic paper on the role of metabolic processes in cancer at length:

Cancer cells originate from normal body cells in two phases. The first phase is the irreversible injuring of respiration. Just as there are many remote causes of plague—heat, insects, rats—but only one common cause, the plague bacillus, there are a great many remote causes of cancer—tar, rays, arsenic, pressure, urethane—but there is only one common cause into which all other causes of cancer merge, the irreversible injuring of respiration. The irreversible injuring of respiration is followed, as the second phase of cancer formation, by a long struggle for existence by the injured cells to maintain their structure, in which a part of the cells perish from lack of energy, while another part succeed in replacing the irretrievably lost respiration energy by fermentation energy. Because of the morphological inferiority of fermentation energy, the highly differentiated body cells are converted by this into undifferentiated cells that grow wildly— the cancer cells. . . . there is today no other explanation for the origin of cancer cells, either special or general. From this point of view, mutation and carcinogenic agent are not alternatives, but empty words, unless metabolically specified. Even more harmful in the struggle against cancer can be the continual discovery of miscellaneous cancer agents and cancer viruses, which, by obscuring the underlying phenomena, may hinder necessary preventive measures and thereby become responsible for cancer cases. (1956, 312)

Warburg is clearly not lacking in ambition: he wished to offer a general theory of carcinogenesis. Not only were cancer cells distinctive in their metabolism, but this metabolic difference was essential to cancer. No other explanation was available, he argued; thus, Warburg proposed his theory (that cancer results from "injuring of respiration") as an alternative to the theory that it is caused by mutations or viruses. This theory was, unfortunately, regarded as relatively implausible by many cancer researchers, until it was revived by work, in part, by Bissell and colleagues in the 1980s. Bissell began her career in cell biology in the 1970s. She came to cancer research as a relative outsider; her graduate research had been in bacteriology and biochemistry. As an outsider, she was less likely to take for granted what others regarded as obvious. Bissell was not the only one. Several other biologists starting in the 1980s and 90s (notably, Soto and Sonnenschein, 1999) began to question the "oncogene paradigm." Together they placed less emphasis on cell-intrinsic mechanisms, and greater emphasis on the role of the tissue microenvironment in cancer. Bissell argued that cancer cells and their microenvironment are in constant interaction over the course of cancer initiation and progression. What at the time was a radical hypothesis has become mainstream: it is now widely accepted that the tumor microenvironment plays a significant role in cancer progression. The causal pathways regulating cell behavior are, in other words, a two-way street. The behavior of both healthy cells and cancer cells is highly context-dependent. The pathway to this insight was not direct.

Bissell's work began with research into cellular metabolism. As a result of this early research, Bissell challenged the idea that intercellular signaling molecules that had previously been regarded as serving mere "housekeeping" functions were not simply permissive conditions, but could be altered in ways that changed cell phenotype. Cells, it turns out, can "change their fate"—once a cell becomes differentiated into a liver, kidney, or skin cell, its fate is not necessarily sealed. This insight had significant implications: it is not simply cell-intrinsic properties—such as mutations to oncogenes and tumor suppressor genes—that determine the typical behaviors of cells, but also the extracellular matrix, the signaling molecules, and molecular milieu.

Bissell drew on the work of developmental biologists, who focus attention on how cells change over time in light of their location, interactions, and structural relationships to surrounding cells and tissues. The possibility that cells could be relatively flexible or less fixed than previously was assumed, was one she drew upon in her research. And she was inspired to work on the role of the tissue microenvironment in part because of several insights coming from developmental biology and some surprising evidence of cancer regression. For instance, Mintz and Illmensee (1975) injected embryonic carcinoma cells into the blastocyst of a mouse and demonstrated that they contributed to the formation of a variety of tissues in cancer-free adult mice. In Kenny and Bissell's words, "[That] otherwise malignant cells could contribute to normal structures provided a striking exposition of the power of tissue context to modify the malignant potential of cancer cells. . . the implications of these experiments, that genetic alterations could be trumped by the microenvironment, were not widely appreciated as the oncogene paradigm and the importance of genetic changes in cancer rapidly took hold" (Kenny and Bissell, 2003, 2).

In one important experiment in 1976, Bissell began investigation of Warburg's thesis: she was able to induce normal rates of glucose transfer in cancer cells by reducing the amount of glucose added to culture and to achieve the reverse effect in normal cells: "I brought tumor cells to the level of normal, and I brought normal to the level of tumor. And, normal cell pattern became tumorigenic, and the cancer cells became normal." (Bissell, 2017, personal interview) had demonstrated that the "constitutive" features of cell were "inducible." That is, the distinctive glucose metabolism of cancer cells could be shifted by shifting the *environment* of the cell.

This was not the only way in which the behavior of a cancer appeared to be influenced by environmental variables. In 1979, one of Bissell's postdoctoral students, Joann Emerman, demonstrated that one could change the shape of a mammary cell by changing the extracellular matrix. Emerman showed that growing the cells in a different type of culture (three-dimensional culture) caused them to maintain milk protein (beta-casein), whereas growing them in a standard culture led to a different phenotype. Another postdoctoral student, Eva Lee, demonstrated that environmental conditions could affect the phenotypic expression of cells in culture, suggesting that cells did not necessarily have "terminal differentiation." In other words, cells simply did not have a fixed phenotype, but could change in response to different environmental conditions. The relationship between cells and their environment was far more dynamic than had been imagined. Moreover, many of the same signals and interactions that shaped development might also play a role in cancer. This was the launching pad for Lee's subsequent work on the possibility of reversing the cancer phenotype by altering the tissue microenvironment.

Bissell's lab was instrumental in demonstrating how and why wounding can induce the growth of tumors via the promotion of molecular mediators such as TGF-I (transforming growth factor). Valerie Weaver, yet another of Bissell's postdoctoral students, and the Bissell lab, developed and modified a three-dimensional extracellular matrix assay to study the progression of human breast cancer in a functionally relevant cell culture model. Essentially, the goal was to develop a tumor environment assay that mirrors the tumor microenvironment histologically.

This same group demonstrated that human breast cancer cells treated with an inhibitory antibody (beta-integrin) could revert the cancer cell to the normal phenotype. That is, signaling molecules associated with tissue structure, the cellular integrins, could promote the assembly of adherens junctions and influence the cytostructure of these cells, suggesting that despite the acquisition of mutations, cancer cells could come to behave like normal cells:

> [D]espite a number of prominent mutations, amplifications, and deletions, signaling events which are linked to the maintenance of normal tissue architecture are sufficient to abrogate malignancy and to repress the tumor phenotype. It is thus fair to state that cellular and tissue architecture act as the most dominant tumor-suppressor of all, and that the phenotype can—and does—override the genotype as long as the tissue architecture is maintained. (Weaver et al., 1997, 243)

Tumor cells treated with the same antibody and injected into nude mice led to a significantly reduced number and size of tumors. Bissell's lab also investigated how the extracellular matrix (ECM) regulates apoptosis and, in particular, which molecules in the extracellular environment promote invasion and metastasis. In particular, one matrix metalloproteinase, MMP3/Str-1 (stromelysin 1), is active in development, as well as during the process of tissue remodeling in the breast and during involution, when ECM remodeling and alveolar regression take place (the process that is initiated after breastfeeding ceases). Such molecules facilitate angiogenesis by breaking down the ECM, but they are also associated with cell proliferation. Bissell's group showed that they have a tumor-promoting effect, and drew a parallel between development and tumor formation that was relatively poorly understood. The molecules are synthesized not by cancer cells but by stromal cells in the tissue microenvironment. In other words, "an altered stromal environment can promote neoplastic formation" (Sternlicht et al., 1997).

Throughout these decades, there was a central theme in Bissell's work and that of her students—namely, that the study of cells required attention to their environment; that intercellular structures and signaling were not merely passive "conditions" for the maintenance of cells ("housekeepers"), but played an active role in cell phenotype and could indeed lead cells to reverse conditions thought to be fixed. In other words, cellular behavior was not intrinsically determined, but context-dependent. They studied the regulatory roles of many extracellular factors. Bissell's close study of development led her to the realization that it was inappropriate to speak of "cell fate," for interactions among signaling molecules and cells could modify the phenotype of a cell over time, and even reverse the course of incipient cancers. This insight was in part a product of Bissell's willingness to traverse disciplinary boundaries. Bissell credits her study of developmental biologists' work with having led her to the insight that contextual interactions are fundamental in creating diverse differentiated tissues.

Soto and Sonnenschein (1987) stumbled on a similar realization when using a new kind of plastic tube to store blood serum in their lab. To their surprise, breast cancer cells were growing unusually fast in an experiment to test the role of inhibiting factors in the blood that were thought to play a role in cancer; typically, if an inhibitor was present in a plate, the cells didn't grow. However, a new plastic resin was causing an estrogen-like substance, *p*-nonylphenol, to leach out of the plastic and to behave like estrogen. In other words, the reason for the unusual growth was the plastic they were storing their samples in. This suggested to Soto and Sonnenschein that many plausible cell-extrinsic factors were at work in the growth of cancer cells; estrogen could be playing an important role in the promotion of cancer. Soto and Sonnenschein spent several decades investigating this puzzle—namely, what role cell-extrinsic factors might play in cancer progression—and eventually proposed the "Tissue Organization Field Theory" (TOFT) in their 1998 book, *The Society of Cells*.

They contrast the TOFT with somatic mutation theory (SMT). According to Soto and Sonnenschein (see, e.g., 1998, 2004), the "default" state of cells is proliferation rather than quiescence. The latter is a view they attribute to mainstream cancer scientists and, in particular, defenders of SMT. Typically, they argue, cancer is caused by loss of control of proliferation, where this control is (ordinarily) maintained by the three-dimensional organization of tissues. In other words, cancer is a problem of tissue organization, not a product of mutations to cancer cells. According to Soto and Sonnenschein, the advantage of this view is that it can explain the regression of cancer and the fact that a cancer cell put into a novel tissue microenvironment can grow into normal, healthy tissue, while SMT cannot. Moreover, they claim that the two views are not simply competing theories, but incommensurable "paradigms" for cancer research. Defenders of SMT presume that proliferation is what requires explaining, whereas defenders of TOFT presume that proliferation is the default state of cells, and so what "requires explaining" is a failure to proliferate, which they attribute to tissue architecture. This is an instance, they claim, of "top-down" causation.[5]

The response to Warburg, Bissell, Soto and Sonnenschein was, at least initially, quite dismissive. The rise of the oncogene paradigm in the 1980s overshadowed much of Warburg's research into cell metabolism, as well as the insights coming from Bissell's and Soto and Sonnenschien's labs on the role of the microenvironment in cancer initiation and progression. However, the view of the cell as in constant dynamic interaction with its intercellular environment eventually became the accepted view among most cancer researchers today. For instance, in 2011, Hanahan and Weinberg revisited their 2000 "Hallmarks of Cancer" in a paper titled "Hallmarks of Cancer: The Next Generation" to include not only "cell-intrinsic" determinants of the cancer phenotype (mutations to oncogenes and tumor suppressor genes), but also cell-*extrinsic* factors. Hanahan and Weinberg explain that "conceptual progress in the last decade has added two emerging hallmarks…reprogramming of energy metabolism and evading immune destruction" (646). They suggest that taking into account the active role of the "tumor microenvironment" "will increasingly affect the development of new means to treat human cancer" (646). Moreover, the work has generated an active research program on the complex interactions between cells, the tissue microenvironment, and the immune system that ordinarily prevent or halt cancer. Arguably, many of these are products of a long history of selection for various mechanisms to suppress cancer.

Cell division in healthy tissue is usually tightly regulated. This is traceable to our evolutionary history. When the first multicellular organisms evolved, cells needed to acquire the capacity to cooperate—that is, limit growth, share resources, and be responsive to intercellular signals. As Amit et al. put it, "To precisely coordinate and integrate cellular decisions such as proliferation,

---

5. For critical discussions, see, e.g., Malaterre (2007) and Bertolaso (2009, 2016). Both nicely illuminate how and why these competing views can be integrated and in what respects they genuinely differ.

differentiation and apoptosis, metazoans developed a set of information relay systems" (2007, 1). Disruption of such systems, or cell signaling pathways associated with growth, is one common feature or hallmark of cancer cells.

Some systems biologists have argued that structural features of networks of signaling pathways within and between cells may make us far less vulnerable to cancer than we might otherwise be. Scale-free networks are networks that have a degree distribution that follows a power law, where a network's degree distribution is the network-wide proportion of edges connected to k other nodes. So scale-free networks have many nodes connected to only one or a few other nodes, and only a few nodes connected to many other nodes (hubs). Scale-free networks have a number of core properties, what systems biologists call "distributed robustness," where "system performance cannot be attributed to any particular element of a system, but rather to a general pattern of how the system elements are connected" (Barrilot et al., 2013, 280). "Robust" systems maintain function despite perturbation, where perturbation might mean elimination or disruption of the activity of nodes. Wagner (2005) has argued that such systems are more likely to evolve.

In scale-free networks, one can eliminate many such nodes without disrupting function. However, if one disrupts several "hubs," then function is destabilized. Thus, scale-free networks are "robust" generally, but are vulnerable to the disruption of highly connected nodes. All things considered, then, scale-free networks may be considered to be relatively "robust" to insult, in comparison with randomly distributed networks. However, once a hub is affected, they can quickly collapse. Such structures may be what we see in cancer, for the "driver genes" in cancer are often hubs in networks that control many core regulatory functions in the cell, including but not limited to advancing the cell cycle (and thus initiating or halting cell division), initiating cell death, and transferring such signals from outside to inside the cell. That is, some of the genes most commonly affected in cancer function as hubs in networks of signaling cascades that affect cell birth and death. So the evolution of scale-free network structure may explain (in part) how we are vulnerable to cancer, as well as why we don't get cancer more often. When many hubs are disrupted, there is destabilization of function (Alon, 2007; Barillot et al., 2013). But, it's also true that this structure and redundancy of some of the same function across different genes and extending into the extracellular environment could protect us from getting cancer more often than we do.

6.4. Concluding Discussion

It is hard to resist the appeal of unified theories; they serve to organize our understanding of the world and give us a research program—that is, problems worth solving, hypotheses to test, and even methods of inquiry. This was very much the case for what Fujimura (1997) has called the "oncogene bandwagon." Consider this abstract for a review paper on the oncogene perspective published by Bishop (1982), shortly before winning the Nobel Prize for Physiology or Medicine:

> Can the cancer cell be understood? Since no one can yet explain how a normal cell controls its growth, it may seem foolhardy to think the abnormal rules governing the growth of a cancer cell can be deciphered. Yet the history of biology records many instances in which the study of abnormalities has illuminated normal life processes. Recent developments in cancer research have added a dramatic new example. For the first time investigators have perceived the dim outline of events that can induce cancerous growth . . .
>
> . . . investigators have been studying these tumor viruses, attempting to define fundamental derangements of the cell that are responsible for cancerous growth. That quest has struck gold.
>
> Although the genes implicated in the development of cancer were first observed in work with viruses, they are not native to the viruses. Indeed, it has turned out that the genes are not even peculiar to cancer cells. They are present and functioning in normal cells as well, and they may be as necessary for the life of the normal cell as they appear to be for the unrestrained growth of a cancer. A final common pathway by which all tumors arise may be part of the genetic dowry of every living cell. (1982, 81)

Bishop's narrative has all the features of a mystery novel: The search for a solution to a problem has so far eluded us. We were looking in the wrong place. Now we have found the one true theory—we have "struck gold" by focusing on the cancer cell and its intrinsic features. And we found the solution in an unexpected place. The shared predisposition to cancer is a feature we all inherit: the shared genetic predisposition. "Our shared genetic dowry" can explain why the search for "tumor viruses" was mistaken. Viruses, Bishop goes on to explain, took up some of the same genes that we all inherit. In mutated forms, these genes in turn lead to cancerous growth in some animals. With this new, unifying theory, we can explain both infectious or viral forms of cancer and an even wider class of facts: no less than all inherited and sporadic forms of cancer across the living world.

The attraction of this story is that (a) the narrative pulls us along, until we suddenly see how, given this unifying story, all the facts about cancer fall into place (Rous sarcoma virus's role in chicken cancers, viral cancers in animals, sporadic forms of cancer, increases of cancer incidence as we age); (b) it gives us a "fundamental" common cause of all cancers; and (c) it gives us clear guidance as to where to search further in our quest to better understand cancer and its causes: in the genes! The power of a unifying narrative is hard to resist. But I think it is better to view the oncogene paradigm as a research program whose success consisted not (exclusively) in what problems it solved, but in yielding several further important puzzles worth exploring.

Several philosophers of science have devoted careful attention various competing "theories of carcinogenesis," considering the limitations and advantages of each.[6] In the above discussion, I took a very different approach. Rather than view the history of cancer research as a history of competing theories, I see it as driven by several, distinct puzzles. In my view, this makes better sense of both the history of cancer research, and the character of scientific explanation in the context of cancer research.

How so? Cancer scientists, apart from their parochial concerns, are largely concerned with how best to intervene to halt disease. The interesting question to ask about cancer research is not "which theory is true," but "which puzzle is most fruitful," where this might be understood in a variety of ways: the puzzle yields more questions worth exploring, or provides us with a way to narrow our focus and identify hypotheses worth testing, or perhaps it provides us with plausible targets for prevention or treatment. Cancer science has moved forward not because scientists are concerned to arrive at the one, best, most unifying theory of cancer, but because they had good questions or puzzles worth exploring.

What makes a puzzle a good one? And what counts as a good solution? Answering these questions, in my view, requires careful attention to historical context. Puzzles change with background knowledge or advances in science. How a puzzle is framed, as well as what we consider a puzzle at all, depends on the state of our current knowledge. New data, technologies, experimental methods, or theoretical advances endow scientists with novel questions, and novel ways of framing answers. The puzzles we might consider worth solving at one point may in short order become mere scaffolding for the next set of puzzles. Science is rather like playing a game where the stakes keep being raised; one may well "win" by the lights of the current generation, but not by the lights of the next generation of players. This has significant import, I think, for philosophical discussions of scientific explanation. With van Fraassen (1980), I think it is well within the norms of colloquial English to say that "Darwin explained evolution." Indeed, arguably what made Darwin's contribution to science so profound is that his work led to, or suggested, many further puzzles requiring solution (Kitcher, 1995). In

---

6. See, e.g., Malaterre, 2007, Marcum, 2005; Bertolaso, 2009. Bertolaso's excellent book (2016) provides a comprehensive and fascinating review of these theories and their history.

my view, we ought to view some of the accomplishments of some of the early cancer researchers similarly.

Many philosophers of science have aimed to provide domain general criteria that mark off successful explanation from, for example, "mere" description, or how possibly from how "actually" explanations (see, e.g., Salmon, 1984). In my view, this goal is rather like the exercise of hoping to give a definitive definition of "species." Species evolve, and sometimes overlap in space and time. We can, for various theoretical or investigative purposes, regard species as "potentially or actually reproductively isolated groups," but this ignores the fact that most species over the history of life on this planet have been asexual (so the criterion of reproductive isolation is technically irrelevant to most species' status); it also occludes the fact that species change over time is gradual, and horizontal gene exchange and cross-species hybrids are far more common than is imagined. While giving us a very precise way of investigating the tips of the branches, this biological species concept in some sense mistakes the forest for the (very recent) trees.

Likewise, in my view, the philosophical work on scientific explanation has focused on the tips of the branches, and so failed to actually capture the rather more complex reality. Scientific explanations are often partial or merely "plausible"—and yet it is exactly these merely plausible stories that move science forward by framing novel puzzles worth exploring. By focusing on the tips of the branches, philosophers ignore the (in my view, far more interesting) tree. Despite the fact that explanation is often treated as a success term in ordinary language, in scientific practice explaining is a process that involves shifting back and forth between different characterizations of the problem to be solved, iterations of more or less adequate solutions, and temporary periods of agreement on a general story, followed by arguments for why this general account doesn't fit the facts. These arguments are then overturned, deepened, or expanded upon by the next generation of scientists.

As we have seen in the above case studies, this has been especially so when facing such complex and heterogeneous phenomena as cancer. Early in the history of cancer research, following the model of infectious disease, the hope was to identify a magic bullet, or single, exclusive, and preferably deterministic cause of cancer. This led first to a search for viral origins of cancer and second to an investigation into the genetic bases of cancer. However, by now it has become clear that there is no single unifying causal story; cancer is not "either" a "viral disease" "or" a "genetic disease." This does not imply, however, that the focus on oncogenes was poor science; it was practical science. Viral and genetic precursors to cancer were important levers for gaining insight into the detailed mechanistic bases for cancer. But it's also true that, as we now know, genes and their products affect and are affected by their cellular milieu, which is in turn affected by the character of immune response, which is in turn affected by the whole organism's environment and its evolutionary and developmental history. So understanding how mutations functioned in the progression of neoplasms provided an anchor for what has become a broader, more integrative and interdisciplinary approach to understanding the progression of the disease (for a discussion, see, e.g., Blassime et al., 2013; Malaterre, 2007).

This account of cancer science, as I've characterized it, requires that we treat explanation by "degrees" and admit distinct "strategies" of explanation. That is, I take these case studies to force us to acknowledge the diversity of legitimate explanatory goals in science. Biologists are not only interested in "why X," but also wish to explain "why possibly," or "how possibly," or "what would we expect if," or "how common, how rare might this be, . . . if we assume circumstances x, y, or z?" This diversity of questions requires a diversity of explanatory strategies. One way in which biologists arrive at answers to many of these hypothetical questions is to build models. Model-building is not a purely hypothetical enterprise in the sense of merely speculative fiction. To build a good model, you need a lot of empirical information about the system(s) of interest you are modeling. Modelers need to have a sense of what conditions are relatively constant and what things vary and under what circumstances.

How do models function in explanation? Mathematical models (absent interpretation) are not propositional, so they cannot stand in inferential relations, as required by Hempel's view that explanations are simply deductive or inductive arguments. However, mathematical models can be interpreted propositionally, as stating something like "laws." In other words, they "say what *would happen* if a certain set of conditions is satisfied by a system," just as laws do (Sober, 2000, 16, emphasis added). So explanations that make appeal to models may be rationally reconstructed such that the "law-models" plus initial conditions constitute deductive arguments. Indeed, in this way, some explanations that appeal to models have something like the structure of deductive nomological explanations.

However, giving such explanations is not—or not centrally—the aim of building models. Models serve in arguments for what *may* happen or what is likely or unlikely to happen, given some set of initial conditions. Such models provide conditions on the possibility of some process or event, or what "would" or "must" happen, provided assumptions of the model hold. Such arguments can function as the "backbone" of any research program, in the following sense. They frame what's relevant or show how and why the relevant factors interact to yield a range of plausible outcomes, provided the system of interest is close (enough) to the system represented by this model. In sum, we can view the explanatory purpose of a theoretical model (such as the multistage theory's models) as giving us tools to think about the causal dynamics of a system of interest, directing a research program, and suggesting hypotheses worth testing. The goal of models such as these is not getting the "causal structure of the world" correct in all its detail, but instead providing a general causal story or a starting point for further elaboration.

In sum, the above studies in puzzle-solving support an "historical" and "iterative" or processual view of explanation. What scientists are doing at any particular point is often not easily demarcated as either discovery or justification (see, e.g., Bechtel and Richardson, 2010), but as an ongoing dynamic process of developing models, theories, and explanatory accounts that range from partial, incomplete, or merely hypothetical to robust and well confirmed by a wide array of evidence. While most philosophers of science would acknowledge this by now, it seems that when we speak about explanation, this fact about practice is forgotten. Explanation is a "success" term, and so, by the lights of most philosophers, it ought to happen after the discovery (and confirmation) process is complete. "Explanation" becomes hypostatized as a thing with a specific logical structure or as referring to a specific kind of fact in the world—a mechanism or a causal structure. But this is to focus on the endpoint, ignoring much of the foundational work along the way.

A central part of current work in philosophy of biology concerns mechanisms: their nature and role in scientific explanation. Clearly the identification of mechanisms is a central part of the biological sciences—particularly molecular biology, genetics, and neuroscience. However, it is by no means unambiguous what counts as a mechanism, and (not coincidentally) philosophers disagree about the extent of emphasis we ought to place on mechanism or mechanistic explanation in the biological sciences. The identification of mechanisms has undoubtedly been a central part of cancer research, in part simply because molecular genetic—largely reductive—approaches to cancer have dominated the past thirty years of work on the disease. But cancer's progression is not strictly determined by cancer cells, but is also governed by the tumor microenvironment, immune response, and tissue architecture. In some ways, the identification of single mechanisms—such as the role of a specific mutation in the regulation of the cell cycle—plays at best a very limited explanatory role in a larger picture of cancer's progression. As we saw in Chapter 3, such mechanisms are highly unstable, local, and context-dependent in their effects.

Thus, while the identification of mechanisms at the cell and molecular levels will always be a central component of cancer research, there are other facts to explain and other dynamics, patterns, and processes of cancer that these mechanisms ought to be situated within. Describing the machinery, causal behavior, or mechanism at work giving rise to any explanandum is, in other words, *only one* of a variety of types of explanation. Scientific explanation is not simply of the facts, but of concepts, patterns, processes, and possibilities. Theoretical or model-based and mechanistic explanations are often two sides of the same coin; both represent the world indirectly and give us important information about the structure of that world. Model-based accounts provide generality and theoretical plausibility arguments; mechanistic evidence can fill in the causal details. Theoretical models and statistical regularities are necessary to tell us which details we ought to seek and where. For instance, as we've seen, evolutionary, dynamical, and "systems" or integrative study networks of interacting causal factors play a role in our understanding of the transition to metastasis, and arguably the mechanistic story does not make sense absent this evolutionary and dynamical perspective.

When explaining any fact about cancer—whether particular events or general patterns, trends, or processes—we are often required to zoom in and out on the object of explanation, appealing to facts at a very small temporal and spatial scale, and then zoom out to consider developmental history, infection, life history, and environment. Different domains of the sciences deploy different methods and tools, focusing on these different temporal and spatial scales. In my view, different explanations capture different kinds of relations—causal dependencies, stable associations, and theoretical generalizations about what one would or should expect in cancer's etiology and progression. That is, explanation can be local, or focus on one very specific local outcome, or it may be theoretical and involve setting out unified "covering" models or theories, or it can require integrative work that links data, models, and explanations across domains. Each are targeted at different explananda: ranging from particular events, to event types, to general patterns and processes.

Moreover, every event causally relevant to some outcome is *not* also explanatorily relevant. Explanation requires selecting those causes that are most relevant to some explanatory target and contrastive (Achinstein, 1985, van Fraassen, 1980); or, perhaps better, most "stable," "specific," and appropriate to a given level of analysis (Woodward, 2010). Knowing what's explanatorily relevant requires a vast body of evidence and background theory: a comprehensive understanding of the relevant science. So in part, this is an empirical matter, and in part a matter of the pragmatics of explanatory discourse. This does not mean, however, that our explanations are not informed by or must not describe the actual causal structure of the world. That is, the norms of explanation are pragmatic, or epistemic, *and* are subject to ontic constraints.

Skyrms (1984) has argued that cause may be an "amiable jumble" or "kludge" concept; our different models of causation capture different aspects of the phenomena under study. So too, perhaps, talk of what it is to "explain" in science is a kludge. Subsumption under laws, for instance, is the provision of modal information of a kind that allows us to make predictions about similar cases and to see classes of phenomena as of a kind. This can be explanatory if our question is: What do the events of this class have in common? Or how are they similarly governed or produced? However, as we have seen, there are many different kinds of requests for explanation: How? Why? Why not? When? Where? How often? What do these have in common? How is this possible? Why is this likely? When is this un-likely? What is likely, given these initial conditions? What is possible, given these initial conditions? Sometimes, answers to such questions are complementary. But it's also true that different types of causal representations of the same system are possible, either at different temporal and spatial scales or with respect to different kinds of outcomes or counterfactual questions, and they may not neatly map onto one another via supervenience or reduction relations, suggesting that there may well be diverse and equally empirically warranted ways of dividing up the causal structure of the world (Godfrey-Smith, 2010; Cartwright, 1999, 2004).

CONCLUSION

Cancer is in fact not one disease but many different diseases. This is one of several reasons why explaining cancer is an iterated (and incomplete) process; while we know more today than thirty or fifty years ago, there is still much that we do not understand. Another reason is that there is so much that we wish to explain: patterns of incidence and mortality, patterns of disease course and responsiveness to therapy of different cancer types and subtypes, and the characteristics of different cancer cells, either in culture or in the body. In explaining all these diverse facts, cancer scientists draw upon experimental and epidemiological evidence, mechanistic models, mathematical models, computer simulations, and genomic, proteomic, and epigenomic data. Each such source of evidence is very particularly relevant to particular outcomes, patterns, and behaviors. It is thus quite difficult to generalize about what it means to "explain cancer," given the diversity of both explanatory practices and, indeed, targets of explanation. Discriminating between adequate and inadequate explanations in cancer science is a highly pragmatic and context-dependent matter.

Are we winning the "war on cancer"? Likewise, this question requires attention to context. There is good reason for optimism, but the results are mixed. Medicine has made enormous strides in the past fifty years. In the 1960s, greater than 90% of children with acute lymphoblastic leukemia (ALL) did not expect to live past five years; 65% of women with multifocal breast cancer did not expect to live past five years. Today, more than 90% in both these groups are expected to survive well into old age. In 1970, 7% of patients diagnosed with lung cancer were still alive after five years; today, 14% are expected to survive. Indeed, for many cancers, it appears that age-adjusted mortality has fallen or leveled off. However, as discussed in Chapter 2, these improvements may in part be illusory, because with more sensitive screening instruments, we may catch some cancers earlier. "Lead-time bias" artificially inflates apparent improvements in mortality. Indeed, one in two men and one in three women in the United States are expected to develop cancer in their lifetimes; and one in four and one in five, respectively, are expected to die from cancer. Cancer is the leading cause of death for all Americans under 85, and the second-leading cause (after heart disease) for those over 85. Moreover, cancer incidence is expected to triple worldwide in the next fifty years, due both to changing demographics in the developing world and to high rates of smoking in China and the Middle East in particular (Parkin, 2001).

Cancer still causes a great deal of suffering; apart from the pain caused by advanced stages of cancer, cancer treatments are some of the most physically devastating medical interventions undertaken in modern medicine. Chemotherapy, radiation, and surgery can take not months but years to recover from, if one is lucky enough to survive them—and some patients never fully recover. Indeed, treatment for cancer itself is so toxic that it can raise the risk of "secondary" cancers: leukemias and lymphomas induced in part by chemotoxic agents and radiation. Targeted "personalized" treatments are on the horizon, but the horizon seems always to be receding.

In short, cancer affects a huge proportion of people in the first world and is anticipated to affect an even larger proportion worldwide. There is much work to do, and not all of this work is purely empirical. There is a wide field of open questions about how theoretical understanding of cancer as a disease informs our practice of cancer prevention and treatment. This book has been concerned primarily with the basic biology of cancer; there are still many open questions about how we move from bench to bedside. While historians and sociologists of medicine have done fascinating work on the social and historical context of cancer research, treatment, and prevention, as well as the social negotiations involved in characterizing the disease, few philosophers of science have devoted serious attention to the nature of cancer as a disease or the complex interface between the science, medical practice, and policies surrounding cancer prevention and treatment (Moss, 2002; Morange, 2011, Bertolaso, 2009; Malaterre, 2007). One of the aims of this book has been to set the stage for further inquiry into the basic biology of cancer and philosophical questions raised about cancer science.

Cancer thus poses a special opportunity and challenge for philosophers of science to be more attentive to how our social context, values, interests, and purposes shape science. Cancer research is by and large directed toward understanding, not only for its own sake, but also for the purposes of treatment and prevention. As such, conceptual and methodological questions about how we understand and investigate cancer cannot be treated in isolation from questions of practical import and values. Social and political context plays an important role in shaping both "external" questions about what gets funded and even strictly "internal" questions in diagnosis and criteria for staging, disease classification, and evidence regarding causation and theory choice.

For good or ill, there is a good deal of money in cancer research. Thus, there is a good deal of jockeying for that money, and what studies get funded, and why, is not determined entirely by intrinsic scientific interest or the potential for benefiting the greatest number at the least cost. In the past twenty-five years, following the passage of the Bayh-Dole Act, there has been a massive shift of funds away from publicly funded basic science, toward privately funded research in biomedical and biotech fields. It is contentious whether the dramatic increase in private investment in biomedical research has improved, or simply changed, the character of biomedical research. Research into costly tests for relatively rare genetic diseases or applications for expensive patented drugs may be financially more remunerative than, for instance, research into cost-effective screening for uninsured or underinsured populations or into the preventative applications of generic drugs. Perhaps not surprisingly, the vast majority of research in private biotech leads to tests and treatments for only those who live in the developed world or simply those who have the means to pay. Cancer pharmaceuticals, radiation, and surgical interventions are some of the most expensive drugs and treatments available. This is particularly concerning given that, over the next fifty years, worldwide cancer incidence is expected to triple; different nations have very different standards of cost control on pharmaceuticals, but many cancer drugs are prohibitively expensive in the developing world. This raises compelling questions about justice and fairness in access to health and healthcare. So, while we have made great strides in cancer medicine, arguably these benefits are far from fairly distributed.

Progress is difficult to measure for other reasons as well. We may have made the most significant advances in surgery, chemotherapy, and radiation in the 1960s and 1970s. Currently, it is much more expensive, time-consuming, and methodologically difficult to make minor improvements in prevention and treatment after the first major improvements have already been made. The layers of required testing for drugs and treatments mean that moving from "bench to bedside" is enormously expensive and can take decades. In very few instances are the promises made on behalf of basic science borne out. It's also probably true, however, that as the science has moved forward, asking new and different questions, and trying out radical ideas, is more difficult, and not simply because of limitations in funding. The more we know and the more institutional inertia there is behind certain ways of studying and thinking about a disease, the more difficult it is to propose genuinely novel research. Perhaps we need to welcome more radical ideas into the fold.

In fact, the question of whether we are winning the "war on cancer" is also somewhat misguided. As is no longer news, Nixon had the wrong metaphor. Cancer is not like many infectious diseases with single causative agents that we can hope to cure or prevent with drugs or immunizations. There is not likely to be a single magic bullet, at least in part because cancer is not one disease. More controversially, there is a sense in which cancer is "inevitable." Trillions of cell divisions occur in the lifetime of an individual, mutations happen, the body's mechanisms of control break down, and so there is always the potential for cancer to come about. Some advocates of evolutionary medicine have argued that cancer is a kind of "evolutionary inevitability"—the product of both "chance and necessity"—mutation and natural selection, at the cellular level (Greaves, 2001). Cellular cooperation on the scale that we see in multicellular organisms like us is a relatively novel achievement in the history of life on earth, and a precarious one. This way of thinking about cancer as a disease may have potential for applications to cancer treatment and prevention. But again, this is only one tool in the arsenal.

I hope to have shown here that philosophers of science might engage with these questions, but I have touched on very few. Indeed, I hope that the audience for this book encompasses not just philosophers, but anyone interested in the opportunity to reflect on cancer from a variety of disciplinary perspectives: the nature of disease and health, causation, explanation and evidence in the biomedical sciences, and, more broadly, science policy and clinical decision-making. I hope that all of us affected by cancer may contribute to this continuing conversation.

THE BASIC BIOLOGY OF CANCER

This appendix offers a broad overview of cancer initiation and progression. Scientific readers may skip this discussion; philosophers with little or no background in the science of cancer may find reading it useful for understanding arguments presented in the book proper. (For a more thorough overview than may be provided here, see Weinberg, 2013; Barillot et al., 2013; Frank, 2007; or Wodarz and Komarova, 2014.)

Cancer is typically characterized as a disease of disorderly cell growth and disruption of tissue organization. Cancer incidence increases by and large with age. Thus, most cancers are called "sporadic" (as opposed to "inherited" or "familial cancer syndromes")—that is, they arise late in life and are largely believed to be due to the gradual acquisition of changes in cells and tissues as we grow older. Incidence curves (a graphical representation, where age is on the x axis and incidence per 10,000 in a given population is on the y axis) for the vast majority of sporadic cancers rise and level off or fall over the course of a lifetime. For instance, breast cancer incidence peaks around age 55–60 and falls off past the age of menopause. However, incidence is in part affected by frequency of screening. Screening early and often leads to higher incidence, as is evidenced by a spike in prostate cancer incidence in the 1980s and 1990s in the United States, following the widespread use of prostate-specific antigen (PSA) screening (Welch and Black, 2010). Early and frequent screening can lead to an inflated appearance of improvements in cancer treatment or improved "disease-free survival" after diagnosis. This is called "lead-time" bias: when the early identification of cancers artificially inflates disease-free survival.

The causes of cancer operate at different temporal and spatial scales. Inherited mutations, environmental carcinogens, viral infection, and an associated long history of inflammation all play causal roles in cancer. The vast majority of cancer risk factors in most parts of the world are environmental—as much as one-third of cancers might be prevented by a reduction or elimination of exposures to environmental carcinogens. Smoking, ultraviolet radiation, infection (human papillomavirus-induced cervical cancer, infection with hepatitis B and C), and lifestyle factors, including diets high in fermented foods, salt, or fat, alcohol consumption, exposure to cytotoxic agents of all sorts in food and the ambient environment, and other environmental exposures all play important causal roles in cancer (the challenges of establishing environmental cancer risk factors in epidemiology are discussed in Chapter 3).

The vast majority of cancers arise in the epithelium—the lining of walls and cavities of the body—including the skin, the throat, and the colon. Cancers of the epithelium are called carcinomas; 80% of cancers are carcinomas. In contrast, sarcomas derive from mesenchymal cell types—fibroblasts (or connective cell tissue types), adipocytes (fat storage cells), osteoblasts (precursors to bone cells), and myocytes (muscle precursors). Leukemias and lymphomas are cancers that arise in blood-forming tissues. Neuroblastomas, gliomas, and glioblastomas are cancers that arise from tissues in the central and peripheral nervous system. Melanomas arise from the pigmented cells of the skin and retina. Cancers' distinctive histopathology (the distinctive shape, structural features, and behavior of cells) can be used to determine the type of tissue in which they arise, but cancers are increasingly classified not only by tissue of origin. They are also subclassified according to their expression of biomarkers, or proteins, their sensitivity to certain promoters, or their particular genetic or genomic features.

Familial patterns of early onset of some cancers identified in the early twentieth century suggested that some cancer types might be due to inherited mutations. For instance, familial adenomatous polyposis (FAP) is a rare inherited susceptibility to the development of polyps in the colon from an early age. Most such polyps are non-malignant; however, over time, individuals with such an inherited susceptibility are more likely to develop malignancies in the colon. Family histories and the construction of Mendelian, or classical, genetic pedigrees help identify dominant or recessive patterns of inheritance. In the case of FAP, dominant patterns of inheritance and linkage analysis led eventually to the identification of genetic markers on the short arm of chromosome 5, and the *APC* gene. *APC* is a "tumor suppressor" gene—a gene that plays an important role in preventing the overgrowth of cells. *APC* produces a protein (also called "Apc") that plays an important causal role in the degradation of beta-catenin, which is an important regulator of cell division. Beta-catenin is expressed at high levels in the interior of the colon crypt, which comprises very small pockets of cells within the colon containing stemlike, or continuously proliferating, cells. The more stemlike cells reside at the very bottom of the crypt, and more differentiated cells appear as one moves up to the surface. At high levels, beta-catenin helps maintain the "stemness" of colon stem cells—it enables them to continue proliferating. These cells proliferate, differentiate, and migrate up to the surface of the colon to form the intestinal lumen— the lining of the gut. Ordinarily, as they migrate up to the surface of the colon, the cells lose their stemness because *APC* triggers the production of a cascade of proteins that eventuate in the degradation of beta-catenin. When the *APC* gene is inactivated, the product (the Apc protein) is not produced, and beta-catenin accumulates at the surface of the crypt, yielding polyps in the colon. Thus, *APC* is a tumor suppressor because it ordinarily prevents the overgrowth of cells, though the gene product has different specific effects in different tissues. Individuals who inherit these mutations have excessive development of polyps, which in turn predisposes them to high rates of colon cancer.

Familial cancer syndromes are relatively rare. However, that some families have a higher cancer incidence at younger ages than other families was one of the main pieces of evidence used to support the "two-hit" theory of cancer (Knudson, 1971). Retinoblastoma was one of the first predictions and confirmations of the "multistage" family of models discussed in Chapter 6, according to which cancer is due to a rate-limited process of the acquisition of mutations. The identification of mutations associated with familial forms of cancer has been important for understanding sporadic or non-familial forms of cancer, because many of the same mutations associated with an increased risk of cancer in families are also found in cancer cells in sporadic cancers. These latter mutations are not inherited, but acquired by the cancer cells during somatic cell division. That is, many of the genes associated with strongly heritable familial forms of cancer (like *TP53* and *APC*) are commonly mutated in sporadic cancers as well. For instance, 50% of sporadic cancers contain mutations to *p53*. These are sometimes called "driver" mutations, because they play such a central role in the regulation of pathways that control when and how cells divide or cease dividing.

Two major classes of mutations are associated with cancer. "Oncogenes," when mutated, cause the overproduction of growth-stimulatory signals. This steady overproduction of growth signals can lead to the proliferation typical of cancer cells. Mutation typically produces gain of function, and such genes are usually dominant. "Tumor suppressor" genes, as we have seen, play the opposite role: they constrain, or "put the brakes on" cellular proliferation; mutations in such genes typically involve a loss of function. The *antigrowth* signals they produce prevent cells from growing without limit. Cells normally receive a signal from their environment, a growth factor that is transduced into the cell by molecules called kinases to signal the activation of transcription and translation of genes that play a role in mitosis.

More than 300 genes have been identified as playing a role in the pathways that are involved in cancer. There are several approaches to identifying such genes: linkage analysis; genome-wide association analysis; and assessing the loss of heterozygosity through polymorphic markers or microarrays. The first oncogenes (*BCR-ABL1, SRC*) were discovered in the late 1970s; the first tumor suppressor genes were identified in the 1980s (*RB, TP53*). In the decades since, many new screening assays for identifying cancer genes have been developed. Recent advances in high-throughput sequencing have led to the identification of hundreds of "cancer genes"—as of 2013, the Catalogue of Somatic Mutations in Cancer (COSMIC) contained 487 genes implicated in cancer (cf. Barillot et al., 2013). In 2006, the National Cancer Institute initiated the Cancer Genome Atlas, which aims at deciphering genomic and epigenomic modifications of tumors in thirty-one different cancers. The "epigenome" includes but is not limited to changes in DNA methylation patterns, histone modification patterns, and other causal factors involved in gene expression, development, and tissue differentiation. The International Cancer

Genome Consortium similarly aims at establishing the molecular profiles of 25,000 tumors from fifty different cancer types.

Cancer cells acquire a variety of capacities—what Hanahan and Weinberg (2000, 2011) have called the "hallmarks" of cancer. Initially they listed six capacities, which are primarily capacities of a cancer cell as it evolves in the multistep process toward a neoplastic state (Hanahan and Weinberg, 2000). The first six hallmarks of cancer are sustained proliferative signaling, evading growth suppressors, resisting cell death, enabling replicative immortality, inducing angiogenesis, and activating invasion and metastasis. Many such hallmarks can be traced to the activity of specific tumor suppressor genes or oncogenes. However, more recently, Hanahan and Weinberg (2011) modified their characterization of cancer to include "emerging hallmarks," such as reprogramming of energy metabolism, evading immune destruction, genetic instability, and inflammation. In other words, cancers are in part the product of causal interactions between cancer cells and the tumor microenvironment.

The vast majority of research on cancer causation for the past twenty-five years has focused on the role of mutations to specific genes in cancer (for a history, see, e.g., Fujimura, 1997; Morange, 1997). However, viruses also play a role in cancer initiation; indeed, some of the first "cancer genes" were discovered as a result of important research into the viral causes of cancer.

Several classes of tumor viruses were identified in the 1960s and 1970s, including HPV (human papillomavirus), which is responsible for most cervical cancers. However, it was still not well understood why or how viral infections led to cancer. One of the most widely believed assumptions in biology in the mid-twentieth century was that information flows only "forward" from DNA to RNA. Temin's theory of viral cancer challenged this assumption; he suggested that information could flow backward from the RSV into the DNA of infected hosts. Temin argued that retroviruses inserted into a host cell might make double-stranded DNA copies of themselves that were then integrated into the host's DNA and replicated when the cell itself replicates via "reverse transcription." Today, reverse transcription is used in the polymerase chain reaction (PCR) technique, gene expression analysis, and cloning and sequencing of RNA. Reverse transcription is the process by which an enzyme mediates the creation of a DNA complement (complementary DNA or cDNA) from an RNA strand.

As discussed in Chapter 6, in 1976, Stehelin, Vogt, Bishop and Varmus found evidence suggesting that the specific gene associated with viral cancer in chickens was not "viral" but was very common, shared across the tree of life. What explained its association with RSV was that the *src* gene had been taken up by RSV in the evolutionary past, and mutated, so as to lead to cancer. The normal form of the gene can *also* be subject to somatic mutations, and thereby contribute to cancer. Varmus and Bishop won the Nobel Prize in 1989 for their discovery.

There are two main classes of "viral" cancers. Each kind of virus produces a different kind of cancer. For instance, HPV, or human papillomavirus, is a DNA virus that leads to warts or papillomas on the skin and also cancer. HPV is a member of a family of papovaviruses that are able to integrate copies of their genomes into host cell DNA. 99% of cervical carcinomas carry fragments of HPV integrated into their DNA. Cells transformed by these and other papovaviruses exhibit characteristics of cancer; they are "immortalized," or proliferate indefinitely in culture, and are anchorage-independent (their growth is not inhibited by neighboring cells).

Of course, the cell cycle is controlled by many factors in interaction, and whether cancer advances to metastasis depends on a variety of extracellular factors. Indeed, as early as the 1980s, it was demonstrated that cancer can be initiated by changes to the tissue microenvironment. Tumors can be produced in RSV-expressing chickens only in the setting of wounding and inflammation (Dolberg et al., 1985). Mammary carcinomas can be induced in mice after implantation of normal epithelial cells into mutagenized mammary fat pads but not when mutagenized epithelial cells are implanted into control fat pads (Maffini et al., 2004).

The extracellular matrix and signals from the tissue microenvironment ordinarily play a suppressive role in preventing cancer, but these same signals can also be hijacked by a cancer or be activated by infection, inflammatory response, and developmental transitions such as those associated with pregnancy. So while the oncogene paradigm has been a useful tool for finding causal pathways leading to cancer, cancer itself is a complex developmental process that involves dynamic reciprocal interactions between the tumor cells and the ecology of the stroma and surrounding tissue in yielding metastasis.

Solid tumors are not simply composed of cancer cells; they are complex tissues containing distinct cell types. The tumor "stroma" is composed of normal cells that play an active role in cancer development. These recruited cells contribute to the six capabilities, listed earlier. Sustained proliferation is one of six of the hallmarks of cancer cells. Cells receive signals from their environment—surrounding cells and tissues—that may cause a cell to divide or cease dividing. The former are called "growth factors." Most growth factors are produced as a result of a series of events in tightly regulated pathways; these signaling pathways can be altered at different points to either stimulate the production of growth factors or increase the sensitivity of cells to these factors. Cancer cells may themselves produce growth factors, which then bind to cognate receptors and cause cells to proliferate. Or cancer cells may send signals to stimulate normal cells within the supporting tumor-associated stroma, which reciprocate by supplying the cancer cells with various growth factors (Cheng et al., 2008; Bhowmick et al., 2004). Finally, the number of receptor proteins displayed at the cancer cell surface can increase, rendering such cells hyper-responsive; the same outcome can result from structural alterations in the receptor molecules themselves. In other words, there are more or less proximate mechanisms for sustained proliferation and more than one way to induce sustained proliferation in a cancer cell.

For solid tumors, the progression to metastasis involves a cascade of processes and a major transition: the epithelial–mesenchymal transition. This transition is essentially the reactivation of a program that all metazoan cells possess, because this is the program that controls the transition from epithelial to mesenchymal cells in development.

The "invasion-metastasis cascade" has seven steps: primary tumor formation, localized invasion, intravasion (or the process of entering the bloodstream), transport through the circulatory system, arrest in microvessels of various organs, extravasion (exiting the bloodstream), and the formation of a micrometastasis and colonization. Most circulating tumor cells fail to found metastases, and most micrometastases fail to develop into metastases or to colonize new tissue microenvironments; this is sometimes called "metastatic inefficiency." Of the successful metastases, only one or a few will seed new macroscopic metastases. This secondary dissemination of metastases is sometimes called a "metastatic shower" because once this capacity to seed new metastases is acquired in a small population, it results in the production of a large number of disseminated metastases.

Many of the steps involved in the transition to metastases involve the reactivation of the extracellular matrix—the transition to a mesenchymal phenotype, which is ordinarily activated during embryogenesis and wound healing. Mesenchymal cells are motile, not fixed, as epithelial cells are, and so can enter the bloodstream and exit into novel environments. This process of transition requires the cooperation of the stroma: normal cells surrounding cancer cells. Cancer cells thus rely upon signals from the surrounding tissue microenvironment to transition to mesenchymal state. The mechanisms involved are very similar to that which would ordinarily be involved in wound healing: the breakdown of the extracellular matrix and the attraction of platelet "couriers" that "disguise" the tumor cell and prevent attacks from the immune system once it enters the bloodstream.

BIBLIOGRAPHY

Abegglen, L. M., Caulin, A. F., Chan, A., Lee, K., et al. (2015). Potential mechanisms for cancer resistance in elephants and comparative cellular response to DNA damage in humans. *JAMA*, *314*(17), 1850–1860.

Achinstein, P. (1964). Models, analogies and theory. *Philosophy of Science*, *31*(4), 328–350.

———. (1984). The pragmatic character of explanation. In *PSA: Proceedings of the Biennial Meeting of the Philosophy of Science Association* (Vol. 2, 275–292). Dordrecht: Reidel.

———. (1991). *Particles and waves: Historical essays in the philosophy of science.* Oxford: Oxford University Press.

Adriaens, P., and DeBlock, A. (2011). *Maladapting minds: Philosophy, psychiatry and evolutionary theory.* Oxford: Oxford University Press.

Aktipis, C. A., Boddy, A. M., Jansen, G., Hibner, U., et al. (2015). Cancer across the tree of life: Cooperation and cheating in multicellularity. *Philosophical Transaction of the Royal Society B*, *370*(1673).

Alberts, B., Johnson, A., Lewis, J., Raff, M., et al. (2002). *Molecular biology of the cell* (4th Ed.). New York: Garland.

Alexandrov, L. B., Ju, Y. S. Haase, K., Van Loo, P., et al. (2016). Mutational signatures associated with tobacco smoking in human cancer. *Science*, *354*(6312), 618–622.

Alon, U. (2007). Network motifs: Theory and experimental approaches. *Nature Reviews: Genetics*, *8*(6), 450–461.

American Joint Committee on Cancer. (2002). Breast. In *AJCC cancer staging manual* (223–240). New York: Springer.

Amit, I., Wides, R., and Yarden, Y. (2007). Evolvable signaling networks of receptor tyrosine kinases: Relevance of robustness to malignancy and to cancer therapy. *Molecular Systems Biology*, *3*(1), 151.

Amundson, R. (2000). Against normal function. *Studies in the History and Philosophy of Biological and Biomedical Sciences*, *31*(1), 33–53.

———. (2005). Disability, ideology, and quality of life: Bias in biomedical ethics. In D. Wasserman, J. Bickenback, and R. Wachbroit (Eds.), *Quality of life and human difference* (101–125). Cambridge: Cambridge University Press.

Anderson, A., Weaver, A., Cummings, P. T., and Quaranta, V. (2006). Tumor morphology and phenotypic evolution driven by selective pressure from the microenvironment. *Cell, 127*(5), 905–915.

Anderson, K., Lutz, C., van Delft, F. W., Bateman, C. M., et al. (2011). Genetic variegation of clonal architecture and propagating cells in leukaemia. *Nature, 469*(7330), 356–361.

Antoniou, A., Pharoah, P. D., Narod, S., Risch, H. A., et al. (2003). Average risks of breast and ovarian cancer associated with BRCA1 or BRCA2 mutations detected in case series unselected for family history: A combined analysis of 22 studies. *The American Journal of Human Genetics, 72*(5), 1117–1130.

Ao, P. (2007). Orders of magnitude change in phenotype rate caused by mutation. *Cellular Oncology: The Official Journal of the International Society for Cellular Oncology, 29*(1), 67.

Ao, P., Galas, D., Hood, L., and Zhu, X. (2008). Cancer as robust intrinsic state of endogenous molecular-cellular network shaped by evolution. *Medical Hypotheses, 70*(3), 678–684.

Arbyn, M., Raifu, A. O., Weiderpass, E., Bay, F., and Antilla, A. (2009). Trends of cervical cancer mortality in the member states of the European Union. *European Journal of Cancer, 45*(15), 2640–2648.

Armitage, P., and Doll, R. (1954). The age distribution of cancer and a multi-stage theory of carcinogenesis. *British Journal of Cancer, 8*(1), 1.

———. (1957). A two-stage theory of carcinogenesis in relation to the age distribution of human cancer. *British Journal of Cancer, 11*(2), 161.

Armstrong, D. M. (1978). *Universals and scientific realism*. Cambridge: Cambridge University Press.

———. (1997). *A world of states of affairs*. Cambridge: Cambridge University Press.

Aronowitz, R. A. (2007). *Unnatural history: Breast cancer and American society*. Chicago: University of Chicago Press.

———. (2010). The converged experience of risk and disease. *Milbank Quarterly, 87*(2), 417–442.

———. (2015). *Risky medicine: Our quest to cure fear and uncertainty*. Chicago: University of Chicago Press.

Ashcroft, R. (2002). What is clinical effectiveness? *Studies in History and Philosophy of Science, Part C: Studies in History and Philosophy of Biological and Biomedical Sciences, 33*(2), 219–233.

Baetu, T. M. (2012). Genes after the human genome project. *Studies in History and Philosophy of Science, Part C: Studies in History and Philosophy of Biological and Biomedical Sciences, 43*(1), 191–201.

Bansal, M., Belcastro, V., Ambesi-Impiombato, A., and Di Bernardo, D. (2007). How to infer gene networks from expression profiles. *Molecular Systems Biology, 3*(1), 78.

Barabasi, A.-L., and Oltvai, Z. N. (2004). Network biology: Understanding the cell's functional organization. *Nature Reviews: Genetics, 5*(2), 101.

Barcellos-Hoff, M. H., Lyden, D., and Wang, T. C. (2013). The evolution of the cancer niche during multistage carcinogenesis. *Nature Reviews: Cancer, 13*(7), 511–518.

Barillot, E., Calzone, L., Hupe, P., Vert, J.-P., and A. Zinovyev. (2013). *Computational systems biology of cancer*. London: CRC Press.

Barry, M. J. (2009). Screening for prostate cancer: The controversy that refuses to die. *New England Journal of Medicine, 360*, 1351–1354.

Batterman, R. W. (2009). Idealization and modeling. *Synthese, 169*(3), 427–446.

Beatty, J. (1995). The evolutionary contingency thesis. In G. Wolter and J. C. Lennox (Eds.), *Concepts, theories, and rationality in the biological sciences* (45–81). Pittsburgh: University of Pittsburgh Press.

Bechtel, W. (2015). Generalizing mechanistic explanations using graph-theoretic representations. In P. A. Braillard and C. Malaterre (Eds.), *Explanation in biology: An enquiry into the diversity of explanatory patterns in the life sciences* (Vol. 11, 199–225). Dordrecht: Springer.

Bechtel, W., and Abrahamsen, A. (2005). Explanation: A mechanist alternative. *Studies in History and Philosophy of Science Part C: Studies in History and Philosophy of Biological and Biomedical Sciences, 36*(2), 421–441.

Bechtel, W., and Richardson, R. C. (2010, originally published 1993). *Discovering complexity: Decomposition and localization as strategies in scientific research*. Boston: MIT Press.

Beral, V., Hannaford, P., and Kay, C. (1988). Oral contraceptive use and malignancies of the genital tract: Results from the Royal College of General Practitioners' oral contraception study. *Lancet, 332*(8624), 13331–13335.

Bertolaso, M. (2009). Towards an integrated view of the neoplastic phenomena in cancer research. *History and Philosophy of the Life Sciences, 31*(1), 79–97.

———. (2016). *Philosophy of cancer: A dynamic and relational view*. Dordrecht: Springer.

Bertolaso, M., Giuliani, A., and Filippi, S. (2013). The mesoscopic level and its epistemological relevance in systems biology. In A. X. C. N. Valente, A. Sarkar, and Y. Gao (Eds.), *Recent advances in systems biology research* (19–36). New York: Nova Science Publishers, Inc.

Beurton, P. J., Falk, R., and Rheinberger, H. J. (Eds.). (2000). *The concept of the gene in development and evolution: Historical and epistemological perspectives*. Cambridge: Cambridge University Press.

Bhowmick, N. A., Neilson, E., and Moses, H. L. (2004). Stromal fibroblasts in cancer initiation and progression. *Nature, 432*(7015), 332–337.

Biddle, J., and Kukla, R. (2017). The geography of epistemic risk. In Kevin C. Elliott and T. Richards (Eds.), *Exploring inductive risk: Case studies of values in science* (215–237). Oxford: Oxford University Press.

Biggs, M. L., Davis, M. D., Eaton, D. L., Weiss, N. S., et al. (2008). Serum organochlorine pesticide residues and risk of testicular germ cell carcinoma: A population-based case-control study. *Cancer Epidemiology, Biomarkers and Prevention, 17*, 2012–2018.

Birch, J. (2017). *The philosophy of social evolution*. Oxford: Oxford University Press.

Bird, A. (2015). The metaphysics of natural kinds. *Synthese*, 1–30.

Bishop, J. M. (1982). Oncogenes. *Scientific American, 246*(3), 80–93.

Bissell, M. (2009). Mina Bissell: Context is everything [an interview by Ben Hhort]. *Journal of Cell Biology, 185*(3), 374–375.

Bissell, M., and Hines, W. (2011). Why don't we get more cancer? A proposed role of the microenvironment in restraining cancer progression. *Nature Medicine, 17*(3), 310–329.

Bissell, M., and LaBarge, M. (2005). Context, tissue plasticity, and cancer: Are tumor stem cells also regulated by the microenvironment? *Cancer Cell, 7*(1), 17–23.

Bissell, M. J. (1981). The differentiated state of normal and malignant cells or how to define a normal cell in culture. *International Review of Cytology, 70*, 27–100.

Bissell, M. J., and Aggeler, J. (1987). Dynamic reciprocity: How do extracellular matrix and hormones direct gene expression? *Progress in Clinical and Biological Research, 249*, 251–262.

Bissell, M. J., Hall, H. G., and Parry, G. (1982). How does the extracellular matrix direct gene expression? *Journal of Theoretical Biology, 99*(1), 31–68.

Bissell, M. (2017) Personal Interview, Berkeley, California.

Blassime, A., Maugeri, P., and Germain, P.-L. (2013). What mechanisms can't do: Explanatory frameworks and the function of the *p53* gene in molecular oncology. *Studies in History and Philosophy of Biological and Biomedical Sciences, 44*, 374–384.

Block, K., Kardana, A., Igarashi, P., and Taylor, H. S. (2000). *In utero* diethylstilbestrol (DES) exposure alters Hox gene expression in the developing Müllerian system. *FASEB Journal, 14*(9), 1101–1108.

Bo, V., and Tucker, A. (2015). Integrating gene regulatory networks to identify cancer-specific genes. *AMIA Joint Summits on Translational Science Proceedings, 2015*, 21–25.

Boorse, C. (1975). On the distinction between disease and illness. *Philosophy and Public Affairs, 5*, 49–68.

———. (1976). What a theory of mental health should be. *Journal for the Theory of Social Behavior, 6*, 61–84.

———. (1977). Health as a theoretical concept. *Philosophy of Science, 44*, 542–573.

———. (1997). A rebuttal on health. In J. Humber and R. Almeder (Eds.), *What is disease?* (1–134). Totowa, NJ: Humana Press.

———. (2014). A second rebuttal on health. *Journal of Medicine and Philosophy, 39*(6), 683–724.

Boveri, T. (1914). English-language edition: 1929. *The origin of malignant tumors*. Baltimore, MD: Williams and Wilkins.

Boyd, R. (1991). Realism, anti-foundationalism and the enthusiasm for natural kinds. *Philosophical Studies, 61*, 127–148.

———. (1999). Homeostasis, species, and higher taxa. In R. A. Wilson (Ed.), *Species: New interdisciplinary essays* (141–185). Cambridge, MA: MIT Press.

Boyle, P., and Brawley, O. W. (2009). Prostate cancer: Current evidence weighs against population screening. *CA: A Cancer Journal for Clinicians, 59,* 220–224.

Braillard, P. A., and Malaterre, C. (2015). *Explanation in biology: An enquiry into the diversity of explanatory patterns in the life sciences* (Vol. 11). Springer, Dordrecht.

Brenner, D. J., Doll, R., Goodhead, D. T., Hall, E. J., et al. (2003). Cancer risks attributable to low doses of ionizing radiation. *Proceedings of the National Academy of Sciences, 100,* 13761–13766.

Bribiescas, R. G., Ellison, P. T., and Gray, P. B. (2012). Male life history, reproductive effort, and the evolution of the genus *Homo*: New directions and perspectives. *Current Anthropology, 53*(S6), S424–S435.

Brigandt, I. (2003). Species pluralism does not imply species eliminativism. *Philosophy of Science, 70,* 1305–1316.

———. (2009). Natural kinds in evolution and systematics: Metaphysical and epistemological considerations. *Acta Biotheoretica, 57*(1–2), 77–97.

Brigandt, I., Green, S., and O'Malley, M. A. (2016). Systems biology and mechanistic explanation. In S. Glennan and P. Illari (Eds.), *The Routledge handbook of mechanisms and mechanical philosophy* (362–375). London: Routledge.

Brigandt, I., and Love, A. (2015). Reductionism in biology. In E. N. Zalta (Ed.), *The Stanford encyclopedia of philosophy* (Fall 2015 Edition). http://plato.stanford.edu/archives/fall2015/entries/reduction-biology/.

Brill, A. B., Tomonaga, M., and Heyssel, R. M. (1962). Leukemia in man following exposure to ionizing radiation: A summary of findings in Hiroshima and Nagasaki and comparison with other human experience. *Annals of Internal Medicine, 56,* 590–609.

Broadbent, A. (2009). Causation and models of disease in epidemiology. *Studies in History and Philosophy of Biological and Biomedical Sciences, 40,* 302–311.

———. (2011a). What could possibly go wrong? A heuristic for predicting population health outcomes of interventions. *Preventive Medicine, 53*(4), 256–259.

———. (2011b). Epidemiological evidence in proof of specific causation. *Legal Theory, 17*: 237–278.

———. (2011c). Causal inference in epidemiology: Mechanisms, black boxes, and contrasts. In P. M. Illari, F. Russo, and J. Williamson (Eds.), *Causality in the sciences* (45–69). Oxford: Oxford University Press.

———. (2012). Philosophy and preventive medicine. *Preventive Medicine, 55*(6), 575–576.

———. (2013). *Philosophy of epidemiology.* London: Palgrave Macmillan.

Brown, M. J. (2014). Values in science beyond underdetermination and inductive risk. *Philosophy of Science, 80*(5), 829–839.

Brown, R. L. (2013). What evolvability really is. *British Journal for the Philosophy of Science, 65*(3), 549–572.

Bruggeman, F. J., and Westerhoff, H. V. (2007). The nature of systems biology. *Trends in Microbiology, 15*(1), 45–50.

Buller, D. (2005). *Adapting minds: Evolutionary psychology and the persistent quest for human nature*. Cambridge, MA: MIT Press/Bradford Books.

Burian, R. M. (1985). On conceptual change in biology: The case of the gene. In D. Depew and B. W. Weber (Eds.), *Evolution at a crossroads: The new biology and the new philosophy of science* (21–42). Cambridge: MIT Press.

———. (2004). Molecular epigenesis, molecular pleiotropy, and molecular gene definitions. *History and Philosophy of the Life Sciences, 26*(1), 59–80.

Burney, L. E. (1959). Smoking and lung cancer: A statement of the Public Health Service. *Journal of the American Medical Association, 171*(13), 1829–1837.

Bursten, J. R. (2016). Smaller than a breadbox: Scale and natural kinds. *British Journal for the Philosophy of Science, 69*(1): 1–23.

Buss, L. (1987). *The evolution of individuality*. Princeton: Princeton University Press.

Cairns, J. (1975). Mutation selection and the natural history of cancer. *Nature, 255*, 197–200.

———. (1978). *Cancer: Science and society*. San Francisco: Freeman and Co.

Cambrosio, A., Keating, P., Mercier, S., Lewison, G., et al. (2006). Mapping the emergence and development of translational cancer research. *European Journal of Cancer, 42*(18), 3140–3148.

Campbell, N. R. ([1920] 2013). *Physics: The elements*. Cambridge: Cambridge University Press.

Caplan, A. L., McCartney, J. J., and Sisti, D. (2004). *Health, disease, and illness: Concepts in medicine*. Washington, DC: Georgetown University Press.

Carnap, R. (1950). *Logical foundations of probability*. Chicago: University of Chicago Press.

Carroll, S. B., Grenier, J. K., and Weatherbee, S. D. (2013). *From DNA to diversity: Molecular genetics and the evolution of animal design*. New York: Wiley.

Carson, R. (1962). *Silent spring*. New York: Houghton Mifflin.

Cartwright, N. (1979). Causal laws and effective strategies. *Noûs, 13*(4), 419–437.

———. (1983). *How the laws of physics lie*. Oxford: Clarendon Press.

———. (1997). Models: The blueprints for laws. *Philosophy of Science, 64*, S292–S303.

———. (1999). *The dappled world: A study of the boundaries of science*. Cambridge: Cambridge University Press.

———. (2004). Causation: One word, many things. *Philosophy of Science, 71*(5), 805–819.

———. (2007). Are RCTs the gold standard? *BioSocieties, 2*(1), 11–20.

Chakravartty, A. (2007). *A metaphysics for scientific realism: Knowing the unobservable*. Cambridge: Cambridge University Press.

———. (2011). Scientific realism and ontological relativity. *Monist, 94*(2), 157–180.

Chalasani, P., and Livingston, R. (2013). Differential chemotherapeutic sensitivity for breast tumors with "BRCAness": A review. *Oncologist, 18*, 909–916.

Chang, H. (2010). The hidden history of phlogiston. *HYLE: International Journal for Philosophy of Chemistry, 16*(2), 47–79.

Charles, D. R., and Luce-Clausen, E. M. (1942). The kinetics of papilloma formation in bendpyrene-treated mice. *Cancer Research*, *2*, 261–263.

Chen, Y., and Hunter, D. (2005). Molecular epidemiology of cancer. *CA: A Cancer Journal for Clinicians*, *55*, 45–55.

Chen, S., and Parmigiani, G. (2007). Meta-analysis of BRCA1 and BRCA2 penetrance. *Journal of Clinical Oncology*, *25*(11), 1329–1333.

Cheng, N., Chytil, A., Shyr, Y., Joly, A., et al. (2008). Transforming growth factor-beta signaling-deficient fibroblasts enhances hepatocyte growth factor signaling in mammary carcinoma cells to promote scattering and invasion. *Molecular Cancer Research*, *6*, 1521–1533.

Chikamatsu. (1931). *Transactions of the Japanese Pathological Society*, *21*, 244.

Clarke, A. E., and Fujimura, J. H. (Eds.). (2014). *The right tools for the job: At work in twentieth-century life sciences*. Princeton, NJ: Princeton University Press.

Clarke, B. (2011). Causality in medicine with particular reference to the viral causation of cancers. Doctoral Dissertation (UCL) University College London.

Cleland, C. E. (2002). Methodological and epistemic differences between historical science and experimental science. *Philosophy of Science*, *69*(3), 447–451.

Cogg, A. (2006). Riddle of infectious dog cancer solved. *New Scientist*. http://www.newscientist.com/article/dn9713-riddle-of-infectious-dog-cancer-solved.html (accessed March 23, 2012).

Comfort, N. (2014). *The science of human perfection: How genes became the heart of American medicine*. New Haven: Yale University Press.

Cooper, R. (2002). Disease. *Studies in History and Philosophy of Biological and Biomedical Sciences*, *33*, 263–282.

Coussens, L. M., and Werb, Z. (2002). Inflammation and cancer. *Nature*, *420*(6917), 860.

Cranor, C. (1997). *Regulating toxic substances: A philosophy of science and the law*. Oxford: Oxford University Press.

———. (2006a). Towards a non-consequentialist approach to acceptable risks. In Tim Lewens (Ed.), *Risk: Philosophical Perspectives* (36–53). London: Routledge.

———. (2006b). *Toxic torts: Science, law, and the possibility of justice*. Cambridge: Cambridge University Press.

———. (2011). *Legally poisoned: How the law puts us at risk from toxicants*. Cambridge, MA: Harvard University Press.

Craver, C. (2007). *Explaining the brain*. Oxford: Oxford University Press.

———. (2009). Mechanisms and natural kinds. *Philosophical Psychology*, *22*(5), 575–594.

———. (2016). The explanatory power of network models. *Philosophy of Science*, *83*(5), 698–709.

Crespi, B., and Summers, K. (2005). Evolutionary biology of cancer. *Trends in Ecology and Evolution*, *20*(10), 545–551.

———. (2006). Positive selection in the evolution of cancer. *Biological Reviews*, *81*(3), 407–424.

Cummins, R. C. (1975). Functional analysis. *Journal of Philosophy, 72*(November), 741–764.

Da Costa, L. F. (2001). Return of de-differentiation: Why cancer is a developmental disease. *Current Opinion in Oncology, 13*(1), 58–62.

Damuth, J., and Heisler, I. L. (1988). Alternative formulations of multilevel selection. *Biology and Philosophy, 3*(4), 407–430.

Darden, L. (2006). *Reasoning in biological discoveries: Essays on mechanisms, interfield relations, and anomaly resolution.* Cambridge: Cambridge University Press.

Deisboeck, T. S., Wang, Z., Macklin, P., and Cristini, V. (2011). Multiscale cancer modeling. *Annual Review of Biomedical Engineering, 13,* 127–155.

Dekeuwer, C. (2015). Defining genetic disease. In P. Huneman, G. Lambert, and M. Silberstein (Eds.), *Classification, disease and evidence* (147–164). Dordrecht, Netherlands: Springer.

de Regt, H. W. (2009). The epistemic value of understanding. *Philosophy of Science, 76*(5), 585–597.

Ding, L., and Wendel, M. (2013). Differences that matter in cancer genomics. *Nature Biotechnology, 31*(10), 892–893.

Dobzhansky, T. (1937). *Genetics and the origin of species* (Vol. 11). New York: Columbia University Press.

———. (1964). Biology, molecular and organismic. *American Naturalist, 4*(4), 443–452.

Dolberg, D. S., Hollingsworth, R., Hertle, M., and Bissell, M. J. (1985). Wounding and its role in RSV-mediated tumor formation. *Science, 230*(4726), 676–678.

Doll, R., and Hill, A. B. (1950). Smoking and carcinoma of the lung: A preliminary report. *British Medical Journal, 2*(4682), 739–748.

———. (1954). The mortality of doctors in relation to their smoking habits: A preliminary report. *British Medical Journal, 328,* 1451–1455.

Doll, R., and Peto, R. (1981). The causes of cancer: Quantitative estimates of avoidable risks of cancer in the United States today. *Journal of the National Cancer Institute, 66,* 1191–1308.

Dong, L., Yu, W.-M., Zheng, H., Loh, M. L., et al. (2016). Leukaemogenic effects of Ptpn11 activating mutations in the stem cell microenvironment. *Nature.* doi: 10.1038/nature20131.

Douglas, H. (2000). Inductive risk and values in science. *Philosophy of Science, 67*(4), 559–579.

———. (2009). *Science, policy, and the value-free ideal.* Pittsburgh: University of Pittsburgh Press.

Dowe, P. (1992). Process causality and asymmetry. *Erkenntnis, 37*(2), 179–196.

Downes, S. M. (2008). Evolutionary psychology. In E. N. Zalta (Ed.), *The Stanford Encyclopedia of Philosophy* (Summer 2014 Edition). http://plato.stanford.edu/archives/sum2014/entries/evolutionary-psychology.

Duhem, P. M. M. (1954). *The aim and structure of physical theory*. Princeton, NJ: Princeton University Press.

Dunn, B. K., and Wickerham, D. L. (2005). Prevention of hormone related cancers: Breast cancer. *Journal of Clinical Oncology*, 23(2), 357–367.

Dunlap, T. R. (1975). *DDT, scientists, citizens, and public policy* (Vol. 2). University of Wisconsin–Madison. (Reissued 2014, Princeton: Princeton University Press.)

Dunlap, T. R. (Ed.). (2008). *DDT, silent spring, and the rise of environmentalism: Classic texts*. Seattle: University of Washington Press.

Dunning, G. M. (1959). Fallout from nuclear tests at the Nevada Test Site No. TID-5551. Division of Biology and Medicine. Radiation Effects of Weapons Branch. UF767.U62 1967a 57-61748 U.S. Federal Civil Defense Administration.

Dupré, J. (1981). Natural kinds and biological taxa. *Philosophical Review*, 90, 66–90.

———. (1984). Probabilistic causality emancipated. *Midwest Studies in Philosophy*, 9(1), 169–175.

———. (1993). *The Disorder of things: Metaphysical foundations of the disunity of science*. Cambridge MA: Harvard University Press.

———. (2012). *Processes of life: Essays in the philosophy of biology*. Oxford: Oxford University Press.

———. (2013). Living causes. In *Aristotelian Society Supplementary Volume* (Vol. 87, No. 1, 19–37). London: Blackwell.

Dvorak, Harold, F. (1986). Tumors: Wounds that do not heal. *New England Journal of Medicine*, 315 (26), 1650–1659.

———. (2015). Tumors: Wounds that do not heal—Redux. *Cancer Immunology Research*, 3(1), 1–11.

Egeblad, M., Nakasone, E., and Werb, Z. (2010). Tumors as organs: Complex tissues that interface with the entire organism. *Developmental Cell*, 18(6), 884–901.

Ekstrand, A. J., James, C. D., Cavenee, W. K., Seliger, B., et al. (1991). Genes for epidermal growth factor receptor, transforming growth factor alpha, and epidermal growth factor and their expression in human gliomas in vivo. *Cancer Research*, 51, 2164–2172.

Elgin, C. Z. (2004). True enough. *Philosophical Issues*, 14(1), 113–131.

Elliott, K. (2011). *Is a little pollution good for you? Incorporating societal values in environmental research*. New York: Oxford University Press.

Ellis, B. (2001). *Scientific essentialism*. Cambridge Studies in Philosophy. Cambridge: Cambridge University Press.

Elmore, J. G., and Fletcher, S. W. (2012). Overdiagnosis in breast cancer screening: Time to tackle an underappreciated harm. *Annals of Internal Medicine*, 156(7), 536–537.

Elston, C. W., and Ellis, I. O. (1991). Pathological prognostic factors in breast cancer. I. The value of histological grade in breast cancer: Experience from a large study with long-term follow-up. *Histopathology*, 19(5), 403–410.

Ereshefsky, M. (Ed.). (1992). *The units of evolution: Essays on the nature of species*. Cambridge: MIT Press.

Ereshefsky, M. (1998). Species pluralism and anti-realism. *Philosophy of Science*, 65(1), 103–120.

———. (2001). *The poverty of the Linnaean hierarchy: A philosophical study of biological taxonomy*. Cambridge: Cambridge University Press.

———. (2009). Defining "health" and "disease." *Studies in the History and Philosophy of Biology and Biomedical Sciences*, 40, 221–227.

Ereshefsky, M., and Pedroso, M. (2013). Biological individuality: The case of biofilms. *Biology & Philosophy*, 28(2), 331–349.

Ereshefsky, M., and Reydon, T. A. (2015). Scientific kinds. *Philosophical Studies*, 172(4), 969–986.

Esserman, L. J., Thompson, I. M., Reid, B., Nelson, P., et al. (2014). Addressing overdiagnosis and overtreatment in cancer: A prescription for change. *Lancet Oncology*, 15(6), e234–e242.

Fagan, M. B. (2012). Waddington redux: Models and explanation in stem cell and systems biology. *Biology & Philosophy*, 27(2), 179–213.

Falk, R. (2000). The gene–A concept in tension. In P. J. Beurton, R. Falk, and H. J. Rheinberger (Eds.), *The concept of the gene in development and evolution: Historical and epistemological perspectives* (317–348). Cambridge: Cambridge University Press.

Fidler, I. (2002). The organ microenvironment and cancer metastasis. *Differentiation*, 70(9–10), 498–505.

———. (2003). The pathogenesis of cancer metastasis: The "seed" and "soil" hypothesis revisited. *Nature Reviews: Cancer*, 3, 1–6.

Fisher, R. A. (1957). Dangers of cigarette smoking. *British Medical Journal*, 2(5039), 297–298.

———. (1958a). Lung cancer and cigarettes? (Letter). *Nature*, 182(4628), 108.

———. (1958b). Cancer and smoking. (Letter). *Nature*, 182 (4635), 596.

Formosa, T. (2011, February 16). *Lecture: Cells, molecules, and cancer class for first year medical students*. Salt Lake City: University of Utah College of Medicine.

Foulds, L. (1954). The experimental study of tumor progression: A review. *Cancer Research*, 14(5), 327–339.

Fox Keller, E. (2000). *The century of the gene*. Cambridge, MA: Harvard University Press.

Frank, S., and Nowak, M. (2004). Problems of somatic mutation and cancer. *BioEssays*, 26(30), 291–299.

Frank, S. A. (2007). *Cancer dynamics: Incidence, inheritance and evolution*. Princeton, NJ: Princeton University Press.

Franklin-Hall, L. R. (2015). Natural kinds as categorical bottlenecks. *Philosophical Studies*, 172(4), 925–948.

Friedman, M. (1974). Explanation and scientific understanding. *Journal of Philosophy*, 71(1), 5–19.

Frigg, R. (2010). Models and fiction. *Synthese*, 172(2), 251–268.

Fujimura, J. (1987). Constructing "do-able" problems in cancer research: Articulating alignment. *Social Studies of Science*, 17(2), 257–293.

———. (1988). The molecular biological bandwagon in cancer research: Where social worlds meet. *Social Problems, 35*(3), 261–283.

———. (1997). *Crafting science: A sociohistory of the quest for the genetics of cancer.* Cambridge, MA: Harvard University Press.

Gannett, L. (1999). What's the cause? The pragmatic dimensions of genetic explanation. *Biology and Philosophy, 14*(3): 349–374.

Garber, D. (1992). *Descartes' metaphysical physics.* Chicago: University of Chicago Press.

Garfinkel, A. (1981). *Forms of explanation.* New Haven, CT: Yale University Press.

Garson, J. (2016). *A critical overview of biological functions.* AG Switzerland: Springer International Publishing.

Garson, J., and Picinnini, G. (2014). Functions must be performed at appropriate rates in appropriate situations. *British Journal for the Philosophy of Science, 65*(1), 1–20.

Gause, G. F. (1966). Aspects of antibiotic research. *Chemistry & Industry, 36,* 1506–1513.

Gavert, N., and Ben-Ze'ev, A. (2010). Coordinating changes in cell adhesion and phenotype during EMT-like processes in cancer. *F1000 Biology Reports, 8*(2), 86.

Gemes, K. (1987). The world in itself: Neither uniform nor physical. *Synthese, 73*(2), 301–318.

Genné-Bacon, E. A. (2014). Thinking evolutionarily about obesity. *Yale Journal of Biology and Medicine, 87*(2), 99.

Gerlinger, M., Horswell, S., Larkin, J., Rowan, A. J., et al. (2014). Genomic architecture and evolution of clear cell renal cell carcinomas defined by multiregion sequencing. *Nature Genetics, 46*(3), 225.

Gerlinger, M., Rowan, A., Horswell, S., Larkin, J., et al. (2012). Intratumor heterogeneity and branched evolution revealed by multiregion sequencing. *New England Journal of Medicine, 366*(10), 883–892.

Germain, P.-L. (2012). Cancer cells and adaptive explanations. *Biology & Philosophy.* http://dx.doi.org/10.1007/s10539-012-9334-2.

Giere, R. N. (1999). *Science without laws.* Chicago: University of Chicago Press.

Gilbert, S., and Eppel, D. (2009). *Ecological developmental biology: Integrating epigenetics, medicine and evolution.* Sunderland: Sinauer Associates.

Gillies, D. (2011). The Russo-Williamson thesis and the question of whether smoking causes heart disease. In P. M. Illari, F. Russo, and J. Williamson, (Eds.), *Causality in the sciences* (110–125). Oxford: Oxford University Press.

Glennan, S. (2002). Rethinking mechanistic explanation. *Philosophy of Science, 69*(S3), S342–S353.

Gluckman, P., Beedle, A., and Hanson, M. (2009). *Principles of evolutionary medicine.* Oxford: Oxford University Press.

Glymour, C. (1980). *Theory and evidence.* Princeton, NJ: Princeton University Press.

Godfrey-Smith, P. (1993). Functions: Consensus without unity. *Pacific Philosophical Quarterly, 74,* 196–208.

———. (1994). A modern history theory of functions. *Noûs, 28*(3), 344–362.

————. (2006). The strategy of model-based science. *Biology and Philosophy*, *21*, 725–740.

————. (2008). Reduction in real life. In J. Hohwy and J. Kallestrup (Eds.), *Being reduced: New essays on reduction, explanation and causation* (52–75). Oxford: Oxford University Press.

————. (2009a). *Darwinian populations and natural selection*. Oxford: Oxford University Press.

————. (2009b). Models and fictions in science. *Philosophical Studies*, *143*(1), 101–116.

————. (2010). Causal pluralism. In H. Beebe, C. Hitchcock, and P. Menzies (Eds.), *The Oxford handbook of causation* (326–337). Oxford: Oxford University Press.

Goodman, N. ([1955] 1983). *Fact, fiction, and forecast*. Cambridge, MA: Harvard University Press.

Gøtzsche, P. C., and Jørgensen, K. J. (2013). Screening for breast cancer with mammography. *Cochrane Database of Systematic Reviews*, 6.

Graham, E. A., Croninger, A. B., and Wynder, E. L. (1957). Experimental production of carcinoma with cigarette tar. IV. Successful experiments with rabbits. *Cancer Research*, *17*(11), 1058–1066.

Greaves, M. (2001). *Cancer: The evolutionary legacy*. Oxford: Oxford University Press.

Greaves, M., and Maley, C. C. (2012). Clonal evolution in cancer. *Nature*, *481*(7381), 306–313.

Green, Sara. (2015). Revisiting generality in biology: Systems biology and the quest for design principles. *Biology & Philosophy*, *30*(5), 629–652.

Green, S., Şerban, M., Scholl, R., Jones, N., Brigandt, I., & Bechtel, W. (2018). Network analyses in systems biology: new strategies for dealing with biological complexity. *Synthese*, 195(4), 1751–1777.

Greene, J. (2007). *Prescribing by numbers: Drugs and the definition of disease*. Baltimore: Johns Hopkins University Press.

Griffiths, P. E. (2001). Genetic information: A metaphor in search of a theory. *Philosophy of Science*, *68*(3), 394–412.

————. (2004). Emotions as natural and normative kinds. *Philosophy of Science*, *71*(5), 901–911.

Griffiths, P., and Stoltz, K. (2013). *Genetics and philosophy: An introduction*. Cambridge: Cambridge University Press.

Griffiths, P. E., and Knight, R. D. (1998). What is the developmentalist challenge? *Philosophy of Science*, *65*, 253–258.

Griffiths, P. E., Pocheville, A., Calcott, B., Stotz, K., et al. (2015). Measuring causal specificity. *Philosophy of Science*, *82*(4), 529–555.

Grivennikov, S. I., Greten, F. R., and Karin, M. (2010). Immunity, inflammation, and cancer. *Cell*, *140*(6), 883–899.

Gupta, P. B., Chaffer, C. L., and Weinberg, R. A. (2009). Cancer stem cells: Mirage or reality? *Nature Medicine*, *15*(9), 1010–1012.

Hacking, I. (1991). A tradition of natural kinds. *Philosophical Studies*, *61*, 109–126.

Hacking, I. (1993). On Kripke's and Goodman's Uses of 'Grue'. *Philosophy*, *68*(265), 269–295. Reprinted In C. Z. Elgin (1997). *The philosophy of Nelson*

Goodman: selected essays, 2. Nelson Goodman's new riddle of induction (211–239). New York: Garland.

———. (1999). *The social construction of what?* Cambridge, MA: Harvard University Press.

———. (2007). Natural kinds: Rosy dawn, scholastic twilight. *Royal Institute of Philosophy Supplements, 61*, 203–239.

Hageman, I. (December 12, 2016). Personal Interview. Washington University in St. Louis, St. Louis, MO.

Hanahan, D., and Coussens, L. (2012). Accessories to the crime: Functions of cells recruited to the tumor microenvironment. *Cancer Cell, 21*, 309–322.

Hanahan, D., and Weinberg, R. A. (2000). The hallmarks of cancer. *Cell, 100*(1), 57–70.

———. (2011). The hallmarks of cancer: The next generation. *Cell, 144*(5), 646–675.

Harman, O., and Dietrich, M. R. (Eds.). (2008). *Rebels, mavericks, and heretics in biology.* New Haven, CT: Yale University Press.

———. (2013). *Outsider scientists: Routes to innovation in biology.* Chicago: University of Chicago Press.

Hart, H., and Honore, T. (2002). *Causation in the law* (2nd Ed.). Oxford: Clarendon Press.

Hatfield, G. (1990). Metaphysics and the new science. In D. Lindberg and R. Westman (Eds.), *Reappraisals of the scientific revolution* (93–166). Cambridge: Cambridge University Press.

Hausman, D. M. (2010). Probabilistic causality and causal generalizations. In E. Eells and J. H. Fetzer (Eds.). *The places of probability in science* (Vol. 284, 47–63). Boston Studies in the Philosophy of Science. Dordrecht: Springer.

———. (2011). Is an overdose of paracetamol bad for one's health? *British Journal for the Philosophy of Science, 62*(3), 657–668.

———. (2012). Health, naturalism, and functional efficiency. *Philosophy of Science, 79*(4), 519–541.

———. (2015). *Valuing health: Well-being, freedom, and suffering.* Oxford: Oxford University Press.

Hawkes, K., O'Connell, J. F., Jones, N. B., Alvarez, H., and Charnov, E. L. (1998). Grandmothering, menopause, and the evolution of human life histories. *Proceedings of the National Academy of Sciences, 95*(3), 1336–1339.

Hawley, K., and Bird, A. (2011). What are natural kinds? *Philosophical Perspectives, 25*, 205–221.

Hecht, S. S. (2002). Cigarette smoking and lung cancer: Chemical mechanisms and approaches to prevention. *Lancet Oncology, 3*, 461–469.

Hecht, S. S., Abbaspour, A., and Hoffman, D. (1988). A study of tobacco carcinogenesis. 42. Bioassay in A/J mice of some structural analogues of tobacco-specific nitrosamines. *Cancer Letters, 42*, 141–145.

Heim, D., Budczies, J., Stenzinger, A., Treue, D., et al. (2014). Cancer beyond organ and tissue specificity: Next-generation-sequencing gene mutation data reveal complex genetic similarities across major cancers. *International Journal of Cancer, 135*(10), 2362–2369.

Heim, E., and Köbele, C. (1995). Spontaneous remission in cancer. *Oncology Research and Treatment, 18*(5), 388–392.

Hempel, C. G. (1965). *Aspects of scientific explanation and other essays in the philosophy of science*. New York: Free Press.

Hempel, C. G., and Oppenheim, P. (1948). Studies in the logic of explanation. *Philosophy of Science, 15*(2), 135–175.

Hendrickson, M. (2011). Exorcizing Schrödinger's ghost: Reflections on "What Is Life?" and its surprising relevance to cancer biology. In H. Gumbrecht, R. Harrison, and M. Henrickson (Eds.), *What is life? The intellectual pertinence of Edwin Schrodinger* (45–105). Stanford, CA: Stanford University Press.

Herget, K. A., Patel, D. P., Hanson, H. A., Sweeney, C., and Lowrance, W. T. (2016). Recent decline in prostate cancer incidence in the United States, by age, stage, and Gleason score. *Cancer Medicine, 5*(1), 136–141.

Hesse, M. B. ([1963] 1966). *Models and analogies in science* (Vol. 36). Notre Dame, IN: University of Notre Dame Press.

Hesslow, G. (1988). The problem of causal selection. In D. J. Hilton (Ed.), *Contemporary science and natural explanation: Commonsense conceptions of causality* (11–32). New York: New York University Press.

Hill, A. B. (1965). The environment and disease: Association or causation? *Proceedings of the Royal Society of Medicine, 58*, 259–300.

Hitchcock, C. (1996a). The role of contrast in causal and explanatory claims. *Synthese, 107*, 395–419.

———. (1996b). Farewell to binary causation. *Canadian Journal of Philosophy, 26*, 335–364.

———. (1998). Causal knowledge: That great guide of human life. *Communication and Cognition, 31*(4), 271–296.

Hitchcock, C., and Knobe, J. (2013). Cause and norm. *Journal of Philosophy, 106*(11), 587–612.

Hoffman, F. O., Apostoaei, A. J., and Thomas, B. A. (2002). A perspective on public concerns about exposure to fallout from the production and testing of nuclear weapons. *Health Physics, 82*(5), 736–748.

Holohan, C., Van Schaeybroeck, S., Longley, D. B., and Johnston, P. G. (2013). Cancer drug resistance: An evolving paradigm. *Nature Reviews: Cancer, 13*(10), 714.

Howick, J. H. (2011). *The philosophy of evidence-based medicine*. New York: Wiley.

Huang, S., Ernberg, I., and Kauffman, S. (2009). Cancer attractors: A systems view of tumors from a gene network dynamics and developmental perspective. *Seminars in Cell & Developmental Biology, 20*(7), 869–876.

Huang, S., and Ingber, D. E. (2007). A non-genetic basis for cancer progression and metastasis: Self-organizing attractors in cell regulatory networks. *Breast Disease, 26*(1), 27–54.

Hughes, R. I. G. (1997). Models and representation. *Philosophy of Science, 64*, S325–S336.

Huneman, Philippe. (2010). Topological explanations and robustness in biological sciences. *Synthese, 177*(2), 213–245.

———. (2017). Outlines of a theory of structural explanations. *Philosophical Studies, 175*(3), 665–702.

Hyman, D. M., Puzanov, I., Subbiah, V., Faris, J. E., et al. (2015). Vemurafenib in multiple nonmelanoma cancers with BRAF V600 mutations. *New England Journal of Medicine, 373*(8), 726–736.

IARC Preamble. (2006). *IARC monographs on the evaluation of carcinogenic risks to humans.* Lyon: World Health Organization, International Agency for Research on Cancer.

Illari, P. (2011). Mechanistic evidence: Disambiguating the Russo-Willliamson thesis. *International Studies in the Philosophy of Science, 25*(2), 139–157.

Illari, P. M., Russo, F., and Williamson, J. (Eds.). (2011). *Causality in the sciences.* Oxford: Oxford University Press.

Ingber Donald, E. (2008). Can cancer be reversed by engineering the tumor microenvironment? *Seminars in Cancer Biology, 18*, 356–364.

Institute of Medicine. (2010). The healthcare imperative: Lowering costs and improving outcomes. Workshop Series Summary.

Institute of Medicine. (2013). *Best care at lower cost: The path to continuously learning health care in America.* Washington, DC: National Academies Press.

Ishimaru, T., Hoshino, T., Ichimaru, M., Okada, H., and Tomiyasu, T. (1971). Leukemia in atomic bomb survivors, Hiroshima and Nagasaki, 1 October 1950–30 September 1966. *Radiation Research, 45*(1), 216–233.

Jacob, F. (1977). Evolution and tinkering. *Science, 196*(4295), 1161–1166.

Jemal, A., Center, M., DeSantis, C., and Ward, E. M. (2010). Global patterns of cancer incidence and mortality rates and trends. *Cancer Epidemiology, Biomarkers, and Prevention, 19*, 1893–1907.

John, S. (2011). Why the prevention paradox is a paradox, and why we should solve it: A philosophical view. *Preventive Medicine, 53*, 250–252.

Jones, S., Chen, W. D., Parmigiani, G., Diehl, F., et al. (2008). Comparative lesion sequencing provides insights into tumor evolution. *Proceedings of the National Academy of Sciences, 105*(11), 4283–4288.

Kant, I. ([1790] 2001). *Critique of the power of judgment* (Trans. P. Guyer). Cambridge: Cambridge University Press.

Kaplan, J. M. (2013). *The limits and lies of human genetic research: Dangers for social policy.* New York: Routledge.

Kauffman, S. A. (1971). Articulation of parts explanations in biology. In R. C. Buck and R. S. Cohen (Eds.), *Boston Studies in the philosophy of science* (Vol. 8, 257–272). Dordrecht: D. Reidel Publishing Company.

Keating, P., and Cambrosio, A. (2003). *Biomedical platforms: Realigning the normal and the pathological in late-twentieth-century medicine.* Cambridge, MA: MIT Press.

———. (2011). *Cancer on trial: Oncology as a new style of practice.* Chicago: University of Chicago Press.

Kemeny, J. G., and Oppenheim, P. (1952). Degree of factual support. *Philosophy of Science, 19*(4), 307–324.

Kennaway, E. L., and Kennaway, N. M. (1947). A further study of the incidence of cancer of the lung and larynx. *British Journal of Cancer, 1,* 260–298.

Kevles, D. J. (1998). *In the name of eugenics: Genetics and the uses of human heredity.* Cambridge: Harvard University Press. Reprint Edition.

Khalidi, M. A. (1998). Natural kinds and crosscutting categories. *Journal of Philosophy, 95*(1), 33–50.

———. (2013). *Natural categories and human kinds: Classification in the natural and social sciences.* Cambridge: Cambridge University Press.

Kincaid, H. (2008). Do we need theory to study disease? Lessons from cancer research and their implications for mental illness. *Perspectives in Biology and Medicine, 51*(3), 367–378.

Kingma, E. (2007). What is it to be healthy? *Analysis, 67,* 128–133.

———. (2010). Paracetamol, poison, and polio: Why Boorse's account of function fails to distinguish health and disease. *British Journal for the Philosophy of Science, 61*(2), 241–264.

———. (2014). Naturalism about health and disease: Adding nuance for progress. *Journal of Medicine and Philosophy: A Forum for Bioethics and Philosophy of Medicine, 39*(6), 590–608.

Kinney, P., and Bissell, M. (2003). Tumor reversion: Correction of malignant behavior by microenvironmental cues. *International Journal of Cancer Research, 107,* 688–695.

Kitano, H. (2004). Biological robustness. *Nature Reviews: Genetics, 5*(11), 826–837.

———. (2007a). Towards a theory of biological robustness. *Molecular Systems Biology, 3*(1), 137.

———. (2007b). The theory of biological robustness and its implication in cancer. In P. Bringmann, E. Butcher, G. Parry, and B. Weiss (Eds.), *Systems biology* (69–88). Berlin: Springer.

Kitcher, P. (1982). Genes. *British Journal for the Philosophy of Science, 33*(4), 337–359.

———. (1992). Gene: Current usages. In E. F. Keller (Ed.), *Keywords in evolutionary biology* (128–131). Boston, MA: Harvard University Press.

———. (1995). *The advancement of science: Science without legend, objectivity without illusions.* Oxford: Oxford University Press.

Kitcher, P., and Salmon, W. (1987). VanFraassen on explanation. *Journal of Philosophy, 84,* 315–330.

Klein, C. A., Blankenstein, T. J., Schmidt-Kittler, O., Petronio, M., et al. (2002). Genetic heterogeneity of single disseminated tumour cells in minimal residual cancer. *Lancet, 360*(9334), 683–689.

Knudson, A. G. (1971). Mutation and cancer: Statistical study of retinoblastoma. *Proceedings of the National Academy of Sciences, 68*(4), 820–823.

Koepsell, T. D., and Weiss, N. (2003). *Epidemiologic methods: Studying the occurrence of illness*. Oxford: Oxford University Press.

Komarova, N. L., Katouli, A., and Wodarz, D. (2009). Combination of two but not three current targeted drugs can improve therapy of chronic myeloid leukemia. *PLOS ONE, 4*(2), e4423–e4424.

Komarova, N. L., and Wodarz, D. (2004). The optimal rate of chromosome loss for the inactivation of tumor suppressor genes in cancer. *Proceedings of the National Academy of Sciences USA, 101*(18), 7017–7021.

———. (2005). Drug resistance in cancer: Principles of emergence and prevention. *Proceedings of the National Academy of Sciences, 102*(7), 9714–9719.

———. (2007). Cancer as a microevolutionary process. In S. Stearns and J. Koella (Eds.), *Evolution in health and disease* (2nd Ed.). (289–299) Oxford: Oxford University Press.

Komarova, N. L., Sadovsky, A. V., and Wan, F. Y. (2008). Selective pressures for and against genetic instability in cancer: A time-dependent problem. *Journal of the Royal Society Interface, 5*(18), 105–121.

Kostić, Daniel. (2016). The topological realization. *Synthese, 195*(1), 79–98.

Kreeger, P. K., and Lauffenburger, D. A. (2009). Cancer systems biology: A network modeling perspective. *Carcinogenesis, 31*(1), 2–8.

Kuhn, T. ([1962] 2012). *The structure of scientific revolutions* (4th Ed.). Chicago: University of Chicago Press.

Kumar, V., Abbas, A. K., Fausto, N., and Aster, J. C. (2010). *Robbins and Cotran pathological basis of disease* (8th Ed.). Philadelphia: Saunders Elsevier.

Lakatos, I. (1980). *The methodology of scientific research programmes* (Vol. 1). *Philosophical Papers*. Cambridge: Cambridge University Press.

La Merrill, M., and Birnbaum, L. S. (2011). Childhood obesity and environmental chemicals. *Mount Sinai Journal of Medicine: A Journal of Translational and Personalized Medicine, 78*(1), 22–48.

La Merrill, M., Emond, C., Kim, M. J., Antignac, J.-P., et al. (2012). Toxicological function of adipose tissue: Focus on persistent organic pollutants. *Environmental Health Perspectives, 121*(2), 162–169.

Landecker, H. (2007). *Culturing life*. Cambridge, MA: Harvard University Press.

Lange, M. (2002). Who's afraid of ceteris-paribus laws? Or: How I learned to stop worrying and love them. *Erkenntnis, 57*(3), 407–423.

———. (2005). Laws and their stability. *Synthese, 144*(3), 415–432.

———. (2007). The end of diseases. *Philosophical Topics, 35*(1–2), 265–292.

———. (2009). *Laws and lawmakers: Science, metaphysics, and the laws of nature*. Oxford University Press.

———. (2013). Really statistical explanations and genetic drift. *Philosophy of Science, 80*(2), 169–188.

———. (2016). *Because without cause: Non-causal explanations in science and mathematics*. Oxford: Oxford University Press.

Lange, M., and Rosenberg, A. (2011). Can there be a priori causal models of natural selection? *Australasian Journal of Philosophy, 89*(4), 591–599.

Laplane, L. (2016). *Cancer stem cells*. Cambridge, MA: Harvard University Press.

Lennox, J. (1995). Health as an objective value. *Journal of Medicine and Philosophy*, *20*(5), 499–511.

Lenski, R. (2011). Evolution in action: A 50,000-generation salute to Darwin. *Microbe*, *6*(1), 30–33.

Leonard, G. D., and Swain, S. M. (2004). Ductal carcinoma in situ, complexities and challenges. *JNCI: Journal of the National Cancer Institute*, *96*(12), 906–920.

Leong, A. S. Y. and Zhuang, Z. (2011). The changing role of pathology in breast cancer diagnosis and treatment. *Pathobiology*, *78*(2), 99–114.

Leuridan, B., and Weber, E. (2011). The IARC and mechanistic evidence. In P. Illari, F. Russo, and J. Williamson (Eds.), *Causality and explanation in the sciences* (91–110). Oxford: Oxford University Press.

Levins, R., and Lewontin, R. (1982). *Dialectics and reductionism in ecology*. Dordrecht: Springer.

Lewens, T. (2007). *Risk: Philosophical perspectives*. Cambridge: Cambridge University Press.

Lewis, D. (1973). Causation. *Journal of Philosophy*, *70*, 556–567.

———. (1983). New work for a theory of universals. *Australasian journal of Philosophy*, *61*(4), 343–377.

Lewis, H., and Esquela-Kerscher, A. (2015). MicroRNA epigenetic systems and cancer. In S. Thiagalingam (Ed.), *Systems Biology of Cancer* (134–153). Cambridge, UK: Cambridge University Press.

Lewontin, R. C. (1970). The units of selection. *Annual Review of Ecological Systems*, *1*, 1–18.

———. (1974). Analysis of variance and analysis of causes. *American Journal of Human Genetics*, *26*(3), 400.

———. (1978). Adaptation. *Scientific American*, *239*(3), 212–218, 220, 222.

———. (1985). Population genetics. In P. J. Greenwood, P. H. Harvey, and M. Slatkin (Eds.), *Evolution: Essays in honor of John Maynard Smith* (3–19). Cambridge: Cambridge University Press.

———. (2001). *The triple helix: Gene, organism, and environment*. Cambridge: Harvard University Press.

Lewontin, R. C., and Levins, R. (1987). *The dialectical biologist*. Cambridge, MA: Harvard University Press.

Liong, M. L. (2012). Blood-based biomarkers of aggressive prostate cancer. *PLOS ONE*, *7*(9), e45802.

Lipinski, K. A., Barber, L. J., Davies, M. N., Ashenden, M., et al. (2016). Cancer evolution and the limits of predictability in precision cancer medicine. *Trends in Cancer*, *2*(1), 49–63.

Liu, K., Love, A., and Travisano, M. (2016). How cancer spreads: Reconceptualizing a disease. In G. Boniolo and M. J. Nathan. (Eds.), *Philosophy of molecular medicine: Foundational issues in research and practice* (100–122). London: Taylor & Francis.

Liu, M., Zhou, J., Chen, Z., and Cheng, A. S. L. (2017). Understanding the epigenetic regulation of tumours and their microenvironments: Opportunities and problems for epigenetic therapy. *Journal of Pathology, 241*(1), 10–24.

Lloyd, E. (1994). Normality and variation: The human genome project and the ideal human type. In C. Cranor (Ed.), *Are genes us?* (99–112). New Brunswick, NJ: Rutgers University Press.

Lloyd, E., and Anderson, C. G. (1993). Empiricism, objectivity, and explanation. *Midwest Studies in Philosophy, 18*(1), 121–131.

Longino, H. E. (1990). *Science as social knowledge: Values and objectivity in scientific inquiry.* Princeton, NJ: Princeton University Press.

Lowe, E. J. (1998). *The possibility of metaphysics: Substance, identity, and time.* New York: Oxford University Press.

Lyon, J. (December 1, 2012). Personal interview. University of Utah, Salt Lake City.

Lyon, J. L., Klauber, M. R., Gardner, J. W., and Udall, K. S. (1979). Childhood leukemias associated with fallout from nuclear testing. *New England Journal of Medicine, 300*(8), 397–402.

Machamer, P., Darden, L., and Craver, C. F. (2000). Thinking about mechanisms. *Philosophy of Science, 67,* 1–25.

Mack, S. C., Witt, H., Piro, R. M., Gu, L., et al. (2014). Epigenomic alterations define lethal CIMP-positive ependymomas of infancy. *Nature, 506,* 445–450.

Maffini, M. V., Soto, A. M., Calabro, J. M., Ucci, A. A., and Sonnenschein, C. (2004). The stroma as a crucial target in rat mammary gland carcinogenesis. *Journal of Cell Science, 117*(8), 1495–1502.

Magnus, D. (2004). The concept of genetic disease. In A. L. Caplan, J. J. McCartney, and D. A. Sisti (Eds.), *Health, disease, and illness: Concepts in medicine* (233–243). Washington, DC: Georgetown University Press.

Magnus, P. D. (2015). John Stuart Mill on taxonomy and natural kinds. *HOPOS: The Journal of the International Society for the History of Philosophy of Science, 5*(2), 269–280.

Malaterre, C. (2007). Organicism and reductionism in cancer research: Toward a systemic approach. *International Studies in the Philosophy of Science, 21*(1), 57–73.

Malkin, D. (2017). Cancer predisposition syndromes 101: A case history and review of the challenges. *University of Toronto Medical Journal, 94*(1), 25.

Marcum, J. A. (2005). Metaphysical presuppositions and scientific practices: Reductionism and organicism in cancer research. *International Studies in the Philosophy of Science, 19*(1), 31–45.

Marmot, M., Altman, D., Cameron, D., et al. (2013). The benefits and harms of breast cancer screening: an independent review. *British Journal of Cancer, 108*(11), 2205–2240.

Martincorena, I., Roshan, A., Gerstung, M., Ellis, P., et al. (2015). High burden and pervasive positive selection of somatic mutations in normal human skin. *Science, 348*(6237), 880–886.

Martin-Ruiz, C., Dickinson, H. O., Keys, B., Rowan, E., et al. (2006). Telomere length predicts poststroke mortality, dementia, and cognitive decline. *Annals of Neurology*, *60*(2), 174–180.

Martins-Green, M., and Bissell, M. J. (1990). Localization of 9E3/CEF-4 in avian tissues: Expression is absent in Rous sarcoma virus-induced tumors but is stimulated by injury. *Journal of Cell Biology*, *110*, 581–595.

Martins-Green, M., and Hanafusa, H. (1997). The 9E3/CEF4 gene and its product the chicken chemotactic and angiogenic factor (cCAF): Potential roles in wound healing and tumor development. *Cytokine & Growth Factor Reviews*, *8*(3), 221–232.

Massimi, M. (2016). Bringing real realism back home: A perspectival slant. In M. Couch and J. Pfeifer (Eds.), *The philosophy of Philip Kitcher* (98–121). Oxford: Oxford University Press.

Maynard Smith, J. (2000a). The concept of information in biology. *Philosophy of Science*, *67*(2), 177–194.

———. (2000b). Reply to commentaries. *Philosophy of Science*, *67*(2), 214–218.

Maynard Smith, J., and Szathmary, E. (1995). *The Major Transitions in Evolution*. Oxford: Oxford University Press.

Mayo, D. (1988). Toward a more objective understanding of the evidence of carcinogenic risk. *Philosophy of Science Association Proceedings*, *2*, 489–503.

Mayo, D., and Hollander, R. (Eds.). (1991). *Acceptable evidence: Science and values in risk management*. New York: Oxford University Press.

Mayr, E. (1982). *The growth of biological thought*. New York: Wiley.

McMullin, E. (1968). What do physical models tell us? *Studies in Logic and the Foundations of Mathematics*, *52*, 385–396.

———. (1985). Galilean idealization. *Studies in History and Philosophy of Science, Part A*, *16*(3), 247–273.

———. (1992). *The inference that makes science*. Aquinas Lectures. Milwaukee, WI: Marquette University Press.

Merlo, L. M. F., Pepper, J. W., Reid, B. J., and Maley, C. C. (2006). Cancer as an evolutionary and ecological process. *Nature Reviews: Cancer*, *6*(12), 924–935.

Michod, R. E. (1997). Evolution of the individual. *The American Naturalist*, *150*(S1), S5–S21.

———. (2000). *Darwinian dynamics: Evolutionary transitions in fitness and individuality*. Princeton: Princeton University Press.

———. (2006). The group covariance effect and fitness trade-offs during evolutionary transitions in individuality. *Proceedings of the National Academy of Sciences*, *103*(24), 9113–9117.

———. (2007). Evolution of individuality during the transition from unicellular to multicellular life. *Proceedings of the National Academy of Sciences*, *104*(Suppl. 1), 8613–8618.

Michor, F., Frank, S. A., May, R. M., Iwasad, Y., and Nowak, M. A. (2003). Somatic selection for and against cancer. *Journal of Theoretical Biology*, *225*, 377–382.

Michor, F., Iwasa, Y., and Nowak, M. A. (2004). Dynamics of cancer progression. *Nature Reviews Cancer*, 4(3), 197–205.

Mill, J. S. (1884). *A system of logic*. London: Longman.

Millikan, R. G. (1999). Historical kinds and the special sciences. *Philosophical Studies*, 95: 45–65.

Millstein, R. L. (2006). Natural selection as a population-level causal process. *British Journal for the Philosophy of Science*, 57(4), 627–653.

Mintz, B., and Illmensee, K. (1975). Normal genetically mosaic mice produced from malignant teratocarcinoma cells. *Proceedings of the National Academy of Sciences USA*, 72, 3585–3589.

Mitchell, S. (2000). Dimensions of scientific law. *Philosophy of Science*, 67(2), 242–265.

———. (2003). *Biological complexity and integrative pluralism*. Cambridge Studies in Philosophy and Biology. Cambridge: Cambridge University Press.

———. (2009). *Unsimple truths: Science, complexity and policy*. Chicago: University of Chicago Press.

Moeller, D. W. ([2005] 2009). *Environmental health* (3rd Ed.). Cambridge, MA: Harvard University Press.

Morabia, A. (2004). *A history of epidemiologic methods and concepts*. New York: Springer.

———. (2011). Until the lab takes it away from epidemiology. *Preventative Medicine*, 53(4–5), 217–220.

Morange, M. (1993). The discovery of cellular oncogenes. *History and Philosophy of the Life Sciences*, 15(1), 45–58.

———. (1997). From the regulatory vision of cancer to the oncogene paradigm. *Journal for the History of Biology*, 30, 1–29.

———. (2007). The field of cancer research: An indicator of present transformations in biology. *Oncogene*, 26, 7607–7610.

———. (2011). History of cancer research. *eLS*. Wiley Online Library. John Wiley and Sons., Ltd. Published Online: November 15, 2011. doi: 10.1002/9780470015902. a0003088.pub2.

———. (2012). What history tells us. 28. What is really new in the current evolutionary theory of cancer? *Journal of Biosciences*, 37(4), 609–612.

Morgan, M. S., and Morrison, M. (Eds.). (1999). *Models as mediators: Perspectives on natural and social science* (Vol. 52). Cambridge: Cambridge University Press.

Mori, H., Colman, S. M., Xiao, Z, Ford, A. M., et al. (2002). Chromosome translocations and covert leukemic clones are generated during normal fetal development. *Proceedings of the National Academy of Sciences USA*, 99, 8242–8247.

Moriyama, I. M. (1978). *Capsule summary of results of radiation studies on Hiroshima and Nagasaki atomic bomb survivors, 1945–75*. Radiation Effects Research Foundation, Hiroshima (Japan).

Morrison, M. (2000). *Unifying scientific theories: Physical concepts and mathematical structures*. Cambridge: Cambridge University Press.

Morgan, M. S., and Morrison, M. (1999). *Models as mediators: Perspectives on natural and social science* (Vol. 52). Cambridge: Cambridge University Press.

Moss, L. (2002). *What genes can't do*. Cambridge, MA: MIT Press.

Mueller, F. (1939). Tabakmissbrauch und Lungencarcinom. *Zeitschrift fur Krebsforschung, 49*, 57–85.

Mueller, M., and Fuesnig, N. E. (2004). Friends or foes: Bipolar effects of the tumor stroma in cancer. *Nature Reviews: Cancer, 4*, 839–849.

Mukherjee, S. (2010). *Emperor of all maladies: A biography of cancer*. New York: Scribner.

Muhoro, L. (July 1, 2017). Personal communication (interview). Washington University in St. Louis, St. Louis, MO.

Murphy, D. (2006). *Psychiatry in the scientific image*. Cambridge, MA: MIT Press.

———. (2008). Health and disease. In A. Plutynski and S. Sarkar (Eds.), *The Blackwell companion to the philosophy of biology* (287–298). Oxford: Blackwell.

Naugler, C. T. (2010). Population genetics of cancer cell clones: Possible implications of cancer stem cells. *Theoretical Biology and Medical Modeling, 7*, 42.

Navin, N., Kendall, J., Troge, J., Andrews, P., et al. (2011). Tumour evolution inferred by single-cell sequencing. *Nature, 472*(7341), 90.

NCI. (2003). Cancer and the environment: What you need to know, What you can do. http://www.cancer.gov/cancertopics/understandingcancer/environment (accessed April 1, 2012).

———. (2011). *The nation's investment in cancer research: An annual plan and budget proposal*. US Department of Health and Human Services, National Cancer Institute.

———. (2014). Genetics of breast and ovarian cancer. URL (no longer available) Current version of similar (updated 2018): https://www.cancer.gov/types/breast/hp/breast-ovarian-genetics/pdq#section/_589 (originally accessed January 16, 2014).

———. (2014). What is cancer? http://www.cancer.gov/cancertopics/cancerlibrary/what-is-cancer (accessed April 1, 2014).

Neander, K. (1991). Functions as selected effects: The conceptual analyst's defense. *Philosophy of Science, 58*(2), 168–184.

Needham, P. (2011). Microessentialism: What is the argument? *Noûs, 45*(1), 1–21.

Nelson, C. M., and Bissell, M. J. (2005). Modeling dynamic reciprocity: Engineering three-dimensional culture models of breast architecture, function, and neoplastic transformation. *Seminars in Cancer Biology, 15*(5), 342–352.

Nelson, C. M., and Bissell, M. J. (2006). Of extracellular matrix, scaffolds, and signaling: Tissue architecture regulates development, homeostasis, and cancer. *Annual Review of Cell Developmental Biology, 22*, 287–309.

Nesse, R. M., and Williams, G. C. (1994). *Why we get sick: The new science of Darwinian medicine*. New York: Vintage Books.

Newbold, R. R. (2011). Developmental exposure to endocrine-disrupting chemicals programs for reproductive tract alterations and obesity later in life. *American Journal of Clinical Nutrition, 94*(6 Suppl.), 1939S–1942S.

NIH. (2009). Environmental factors in cancer: Reducing environmental cancer risk, what we can do now. President's Commission Report. http://deainfo.nci.nih.gov/advisory/pcp/annualReports/ (accessed June 6, 2014).

Nguyen, L. V., Vanner, R., Dirks, P., and Eaves, C. J. (2010). Cancer stem cells: An evolving concept. *Nature Reviews Cancer, 12*(2), 220–228.

Niven, W. D. (Ed.). (1965). *The scientific papers of James Clerk Maxell.* New York: Dover.

Nordling, C. O. (1953). A new theory on the cancer-inducing mechanism. *British Journal of Cancer, 7*(1), 68.

Nowak, M., Michor, F., Komarova, N., and Iwasa, Y. (2004). Evolutionary dynamics of tumor suppressor gene inactivation. *Proceedings of the National Academy of Sciences, 101*(29), 10635–10638.

Nowak, M. A. (2006). *Evolutionary dynamics: Exploring the equations of life.* Cambridge, MA: Belknap Press/Harvard University Press.

Nowell, P. C. (1976). The clonal evolution of tumor cell populations. *Science, 194,* 23–28.

Odling-Smee, F. J., Laland, K. N., and Feldman, M. W. (2003). *Niche construction: The neglected process in evolution.* Princeton, NJ: Princeton University Press.

Okasha, S. (2002). Darwinian metaphysics: Species and the question of essentialism. *Synthese, 131*(2), 191–213.

———. (2005). Multi-level selection and the major transitions in evolution. *Philosophy of Science Proceedings, 72,* 1013–1025.

———. (2006a). *Evolution and the levels of selection.* Oxford: Oxford University Press.

———. (2006b). Multi-level selection and the major transitions in evolution. *Philosophy of Science Proceedings, 72,* 1013–1025.

Okruhlik, K. (1994). Gender and the biological sciences. *Canadian Journal of Philosophy,* Supplementary, *20,* 21–42.

Olumi, A. F., Grossfeld, G. D., Hayward, S. W., Carroll, P. R., et al. (1999). Carcinoma-associated fibroblasts direct tumor progression of initiated human prostatic epithelium. *Cancer Research, 59*(19), 5002–5011.

O'Malley, M. A., and Dupré, J. (2005). Fundamental issues in systems biology. *BioEssays, 27*(12), 1270–1276.

Oreskes, N., and Conway, P. (2010). *Merchants of doubt: How a handful of scientists obscured the truth on issues from tobacco smoke to global warming.* New York: Bloomsbury Press.

Orton, N. C., Innes, A. M., Chudley, A. E., and Bech-Hansen, N. T. (2008). Unique disease heritage of the Dutch-German Mennonite population. *American Journal of Medical Genetics Part A, 146*(8), 1072–1087.

Oyama, S. (2000). *The ontogeny of information* (2nd Ed.). Durham, NC: Duke University Press.

Özdemir, B. C., Pentcheva-Hoang, T., Carstens, J. L., Zheng, X., et al. (2014). Depletion of carcinoma-associated fibroblasts and fibrosis induces immunosuppression and accelerates pancreas cancer with reduced survival. *Cancer Cell, 25*(6), 719–734.

Parker-Pope, T. (2013). Scientists seek to rein in diagnoses of cancer. *New York Times*, July 29.

Parkin, D. M. (2001). Global cancer statistics in the year 2000. *Lancet Oncology, 2*(9), 533–543.

Paul, D. (1995). *Controlling human heredity: 1865 to the present (control of nature)*. New York: Humanity Books. Reprint edition.

Pepper, J. W., Sprouffske, K., and Maley, C. C. (2007). Animal cell differentiation patterns suppress somatic evolution. *PLOS Computational Biology, 3*(12), e250.

Perera, F. P. (1995). Molecular epidemiology and the prevention of cancer. *Environmental Health Perspectives, 103*(Suppl. 8), 233–236.

Pierce, G. B., Shikes, R. H., and Fink, L. M. (1978). *Cancer: A problem of developmental biology*. Englewood Cliffs, NJ: Prentice Hall.

Pigliucci, M. (2001). *Phenotypic plasticity: Beyond nature and nurture*. Baltimore: Johns Hopkins University Press.

———. (2008). Is evolvability evolvable? *Nature Reviews: Genetics, 9*(1), 75.

Plutynski, A. (2013). Cancer and the goals of integration. *Studies in History and Philosophy of Science, Part C: Studies in History and Philosophy of Biological and Biomedical Sciences, 44*(4), 466–476.

Preston, D., Kusumi, S., Tomonaga, M., Izumi, S., et al. (1994). Cancer incidence in atomic bomb survivors. Part 3: Leukemia, lymphoma and multiple myeloma, 1950–1987. *Radiation Research, 137*, S68–S97.

Proctor, R. (1996). *Cancer wars: How politics shapes what we do and do not know about cancer*. New York: Basic Books.

Provenzano, P. P., Cuevas, C., Chang, A. E., Goel, V. K., et al. (2012). Enzymatic targeting of the stroma ablates physical barriers to treatment of pancreatic ductal adenocarcinoma. *Cancer Cell, 21*(3), 418–429.

Puliti, D., Duffy, S. W., Miccinesi, G., De Koning, H., et al. (2012). Overdiagnosis in mammographic screening for breast cancer in Europe: A literature review. *Journal of Medical Screening, 19*(Suppl. 1), 42–56.

Putnam, H. (1973). Meaning and reference. *Journal of Philosophy, 70*(19), 699–711.

Quammen, M. (2008). Contagious cancer: The evolution of a killer. *Harpers*. http://harpers.org/archive/2008/04/0081988 (accessed March 23, 2012).

Quanstrum, K., and Hayward, R. A. (2010). Lessons from the mammography wars. *New England Journal of Medicine, 363*, 1076–1079.

Queller, D. C., and Strassmann, J. E. (2009). Beyond society: The evolution of organismality. *Philosophical Transactions of the Royal Society B: Biological Sciences, 364*(1533), 3143–3155.

Railton, P. (1978). A deductive-nomological model of probabilistic explanation. *Philosophy of Science Association, 45*(2), 206–226.

———. (1981). Probability, explanation and information. *Synthese, 48*, 233–256.

Rakoff-Nahoum, S. (2006). Why cancer and inflammation? *Yale Journal of Biology and Medicine, 79*(3–4), 123–130.

Redfield, R. J. (2002). Is quorum sensing a side effect of diffusion sensing? *Trends Microbiology*, *10*, 365–370.

Reid, B. J., Li, X., Galipeau, P. C., and Vaughan, T. L. (2010). Barrett's oesophagus and oesophageal adenocarcinoma: Time for a new synthesis. *Nature Review: Cancer*, *10*, 87–101.

Reisch, G. A. (2016). Aristotle in the cold war: On the origins of Thomas Kuhn's *The Structure of Scientific Revolutions'*. In R. J. Richards and L. Daston (Eds.), *Kuhn's "Structure of Scientific Revolutions" at fifty: Reflections on a science classic* (12–31). Chicago: University of Chicago Press.

Reisman, K., and Forber, P. (2005). Manipulation and the causes of evolution. *Philosophy of Science*, *72*(5), 1113–1123.

Reiss, J. (2015a). A pragmatist theory of evidence. *Philosophy of Science*, *82*(3), 341–362.

———. (2015b). *Causation, evidence, and inference*. London: Routledge.

Reya, T., Morrison, S. J., Clarke, M. F., and Weissman, I. L. (2001). Stem cells, cancer, and cancer stem cells. *Nature*, *414*(6859), 105.

Rode, A., Maass, K. K., Willmund, K. V., Lichter, P., and Ernst, A. (2016). Chromothripsis in cancer cells: An update. *International Journal of Cancer*, *138*(10), 2322–2333.

Rose, G. (2008). *Rose's strategy of preventive medicine* (K.-T. Khaw and M. Marmot, Eds.). Oxford: Oxford University Press.

Rosenblueth, A., & Wiener, N. (1945). The role of models in science. *Philosophy of science*, *12*(4), 316–321.

Ross, L. N., and Woodward, J. F. (2016). Koch's postulates: An interventionist perspective. *Studies in History and Philosophy of Science, Part C: Studies in History and Philosophy of Biological and Biomedical Sciences*, *59*, 35–46.

Rothman, K. (1976). Causes. *Journal of Epidemiology*, *104*(6), 587–592.

Rothman, K., and Greenland, S. (1998). *Modern epidemiology*. Philadelphia: Lippincott.

———. (2005). Causation and causal inference in epidemiology. *American Journal of Public Health*, *95*, 144–150.

Rouse, J. (2001). Two concepts of practices. In T. R. Schatzki, K. Knorr-Cetina, and E. von Savigny (Eds.), *The practice turn in contemporary theory* (189–198). London: Routledge.

Royall, R. (1997). *Statistical evidence: A likelihood paradigm* (Vol. 71). London: CRC Press.

Russo, F., and Williamson, J. (2007). Interpreting causality in the health sciences. *International Studies in the Philosophy of Science*, *21*(2), 157–170.

Sakr, W. A., Haas, G. P., Cassin, B. F., Pontes, J. E., and Crissman, J. D. (1993). The frequency of carcinoma and intraepithelial neoplasia of the prostate in young male patients. *Journal of Urology*, *150*, 379–385.

Salama, M. (2012, January 24). Email communication, following upon his lecture on cancer pathology, University of Utah Medical School.

Salmon, W. (1971). *Statistical explanation and statistical relevance*. Pittsburgh: Pittsburgh University Press.

———. (1984). *Scientific explanation and the causal structure of the world*. Princeton, NJ: Princeton University Press.

———. (1990). Scientific explanation: Causation and unification. *Critica: Revista Hispanoamericana de Filosofia, 2*(66), 3–23.

Sarkar, S. (1996). Biological information: A skeptical look at some central dogmas of molecular biology. In S. Sarkar (Ed.), *The philosophy and history of molecular biology: New perspectives* (187–231). Dordrecht: Kluwer.

———. (2005). *Biodiversity and environmental philosophy: An introduction.* Cambridge University Press. (2005).

Schaffner, K. (1993). *Discovery and explanation in biology and medicine.* Chicago: University of Chicago Press.

———. (1997). Causing harm: Epidemiological and physiological concepts of causation. In D. Mayo and R. Hollander (Eds.), *Acceptable risk: Science and values in risk management* (204–218). Oxford: Oxford University Press.

Schank, J. C., and Wimsatt, W. C. (1986). Generative entrenchment and evolution. In *PSA: Proceedings of the Biennial Meeting of the Philosophy of Science Association* (Vol. 1986, No. 2, 33–60). Dordrecht: D. Reidel.

Schedin, P. (2006). Pregnancy-associated breast cancer and metastasis. *Nature Reviews: Cancer, 6,* 281–291.

Schnitt, S. (2010). Local outcomes in ductal carcinoma in situ based on patient and tumor characteristics. *Journal of the National Cancer Institute Monographs, 2010*(41), 158–161.

Schröder, F. H., Hugosson, J., Roobol, M. J., Tammela, T., et al. (2009). Screening and prostate-cancer mortality in a randomized European study. *New England Journal of Medicine, 360,* 1320–1328.

Schupbach, J. N. (2017). Experimental explication. *Philosophy and Phenomenological Research, 94*(3), 672–710.

———. (2018). Robustness analysis as explanatory reasoning. *British Journal for the Philosophy of Science, 69*(1), 275–300.

Schwartz, P. H. (2007). Defining dysfunction: Natural selection, design, and drawing a line. *Philosophy of Science, 74*(3), 364–385.

———. (2008). Risk and disease. *Perspectives in Biology and Medicine, 51*(3), 320–334.

———. (2014a). Small tumors as risk factors not disease. *Proceedings of the 2012 Biennial Meeting of the Philosophy of Science Association, Part 2: Symposia Papers.*

———. (2014b). Reframing the disease debate and defending the biostatistical theory. *Journal of Medicine and Philosophy, 39*(6), 572–589.

Scolyer, R. A., Murali, R., McCarthy, S. W., and Thompson, J. F. (2010). Histologically ambiguous ("borderline") primary cutaneous melanocytic tumors: approaches to patient management including the roles of molecular testing and sentinel lymph node biopsy. *Archives of Pathology & Laboratory Medicine, 134*(12), 1770–1777.

Sekido, Y., Fong, K. M., and Minna, J. D. (2003). Molecular genetics of lung cancer. *Annual Review of Medicine, 54,* 73–87.

SEER Database. Prostate Cancer Fact Sheet. http://seer.cancer.gov/statfacts/html/prost.html (accessed June 1, 2014).

SEER Database. Surveillance, Epidemiology and End Results Program. Annual Report to the Nation on the Status of Cancer. Racial Disparities in Cancer Incidence and Mortality (accessed June 1, 2014).

Senkus, E., Kyriakides, S., Ohno, S., Penault-Llorca, F., et al. (2015). Primary breast cancer: ESMO clinical practice guidelines for diagnosis, treatment and follow-up. *Annals of Oncology, 26*(Suppl. 5), v8–v30.

Seyfried, T. N., and Shelton, L. M. (2010). Cancer as a metabolic disease. *Nutrition & Metabolism, 7*(1), 7.

Shapiro, L., and Sober, E. (2012). Against proportionality. *Analysis, 72*(1), 89–93.

Shea, N., Pen, I., and Uller, T. (2011). Three epigenetic information channels and their different roles in evolution. *Journal of Evolutionary Biology, 24*(6), 1178–1187.

Shrader-Frechette, K. (2007). *Taking action, saving lives: Our duties to protect environmental and public health*. Oxford: Oxford University Press.

Siegel, R. L., Miller, K. D., and Jemal, A. (2016). Cancer statistics, 2016. *CA: A Cancer Journal For Clinicians, 66*(1), 7–30.

Simon, H. A. (1962). The architecture of complexity. Reprinted in Herbert A. Simon, *The sciences of the artificial* (84–118), Cambridge, MA: MIT Press, 1969.

Skloot, Rebecca. (2010). *The immortal life of Henrietta Lacks*. New York: Broadway Books.

Skrabanek, P. (1992). The poverty of epidemiology. *Perspectives on Biology and Medicine, 35*(2), 182–185.

———. (1994). The emptiness of the black box. *Epidemiology, 5*(5), 553–555.

Skyrms, Brian. (1980). *Causal necessity*. New Haven, CT: Yale University Press.

———. (1984). EPR: lessons for metaphysics. In P. French, T. Uehling, and H. Wettstein (Eds.), *Midwest studies in philosophy IX* (245–255). Minneapolis: University of Minnesota Press.

Slamon, D. J., Clark, G. M., Wong, S. G., Levin, W. J., et al. (1987). Human breast cancer: Correlation of relapse and survival with amplification of the HER-2/neu oncogene. *Science, 235*, 177–182.

Slater, M. H. (2014). Natural kindness. *British Journal for the Philosophy of Science*, axt033.

Smallbone, K., Gatenby, R. A., Gillies, R., Maini, P. K., and Gavaghan, D. J. (2007). Metabolic changes during carcinogenesis: Potential impact on invasiveness. *Journal of Theoretical Biology, 244*, 703–713.

Smith, K. (2007). Toward an adequate account of genetic disease. In H. Kincaid and J. McKitrick (Eds.), *Establishing medical reality: Essays in the metaphysics and epistemology of biomedical science* (83–110). New York: Springer.

Snyder, L. J. (2006). *Reforming philosophy*. Chicago: University of Chicago Press.

Sober, E. (1980). Evolution, population thinking, and essentialism. *Philosophy of Science, 47*(3), 350–383.

———. (2000). *Philosophy of biology* (2nd Ed.). Boulder, CO: Westview Press.

———. (2001). The meaning of genetic causation. In A. Buchanan, D. W. Brock, N. Daniels, and D. Wikler (Eds.), *From chance to choice: Genetics and justice* (347–369). Cambridge: Cambridge University Press.

———. (2008). *Evidence and evolution: The logic behind the science.* Cambridge University Press.

———. (2011). A priori causal models of natural selection. *Australasian Journal of Philosophy, 89*(4), 571–589.

Sober, E., and Wilson, D. S. (1999). *Unto others: The evolution and psychology of unselfish behaviour.* Harvard University Press.

Solomon, M. (2011). Just a paradigm: Evidence-based medicine in epistemological context. *European Journal for Philosophy of Science, 1*(3), 451–466.

———. (2015). *Making medical knowledge.* New York: Oxford University Press.

Sonnenschein, C., and Soto, A. M. (1998). An updated review of environmental estrogen and androgen mimics and antagonists. *Journal of Steroid Biochemistry and Molecular Biology, 65*(1–6), 143–150.

———. (1998). *Society of cells: Cancer and control of cell proliferation.* New York: Garland Science.

———. (2000). Somatic mutation theory of carcinogenesis: Why it should be dropped and replaced. *Molecular Carcinogenesis, 29*(4), 205–211.

———. (2013). The aging of the 2000 and 2011 "Hallmarks of Cancer" reviews: A critique. *Journal of Biosciences, 38*(3), 651–663.

Soto, A. M., and Sonnenschein, C. (1987). Cell proliferation of estrogen-sensitive cells: The case for negative control. *Endocrine Reviews, 8*(1), 44–52.

———. (2004). The somatic mutation theory of cancer: Growing problems with the paradigm? *Bioessays, 26,* 1097–1107.

Sorensen, R. (2011). Para-natural kinds. In J. K. Campbell, M. O'Rourke, and M. Slater (Eds.), *Carving nature at its joints: Natural kinds in metaphysics and science* (113–128). Topics in Contemporary Philosophy (Vol. 8). Cambridge, MA: Bradford Books.

Spencer, S. L., Gerety, R. A., Pienta, K. J., and Forrest, S. (2006). Modeling somatic evolution in tumorigenesis. *PLOS Computer Biology, 2*(8), e108.

Spirtes, P., Glymour, C. N., and Scheines, R. (2000). *Causation, prediction, and search.* Cambridge, MA: MIT press.

Stanford, P. K. (2010). *Exceeding our grasp: Science, history, and the problem of unconceived alternatives.* Oxford: Oxford University Press.

———. (2017). So long and thanks for all the fish: Metaphysics and the philosophy of science. In M. Slater (Ed.), *Metaphysics and the philosophy of science: New essays* (127–141). Oxford: Oxford University Press.

Stankov, A., Bargallo-Rocha, J. E., Silvio, A., Ramirez, M. T., et al. (2012). Prognostic factors and recurrence in breast cancer: Experience at the National Cancer Institute of Mexico. *ISRN Oncology, 2012.*

Stearns, S. C., and Ebert, D. (2001). Evolution in health and disease: Work in progress. *Quarterly Review of Biology, 76,* 417–432.

Stearns, S. C., and Koella, J. (Eds.). (2008). *Evolution in health and disease.* New York: Oxford University Press.

Steel, D. (2007). *Across the boundaries: Extrapolation in biology and social science.* Oxford: Oxford University Press.

———. (2014). *Philosophy and the precautionary principle.* Cambridge: Cambridge University Press.

Stehelin, D., Varmus, H. E., Bishop, J. M., and Vogt, P. K. (1976). DNA related to the transforming gene(s) of avian sarcoma viruses is present in normal avian DNA. *Nature, 260*(5547), 170–173.

Stephens, C. (2004). Selection, drift, and the "forces" of evolution. *Philosophy of Science, 71*(4), 550–570.

Stephens, P. J., Greenman, C. D., Fu, B., Yang, F., et al. (2011). Massive genomic rearrangement acquired in a single catastrophic event during cancer development. *Cell, 144*, 27–40.

Sterelny, K., and Griffiths, P. E. ([1999] 2012). *Sex and death: An introduction to philosophy of biology.* Chicago: University of Chicago Press.

Sternlicht, M. D., Lochter, A., Sympson, C. J., Huey, B., et al. (1999). The stromal proteinase MMP3/stromelysin-1 promotes mammary carcinogenesis. *Cell, 98*, 137–146.

Stocks, P. (1947). *Studies on medical and population subjects: No. 1, Regional and local differences in cancer death rates.* London: HMSO.

———. (1953). A study of the age curve for cancer of the stomach in connection with a theory of the cancer producing mechanism. *British Journal of Cancer, 7*, 407–427.

Strevens, M. (2008). *Depth: An Account of scientific explanation.* Cambridge, MA: Harvard University Press.

Stricker, T., and Kumar, V. (2010). Neoplasia. In V. Kumar, A. Abbas, N. Fausto, and J. Aster (Eds.), *Robbins and Cotran's Pathological basis of disease* (259–331). Philadelphia: Saunders Elsevier.

Sturtevant, A. H. (1913). The linear arrangement of six sex-linked factors in *Drosophila*, as shown by their mode of association. *Journal of Experimental Zoology, 14*(1), 43–59.

Sugiura, K. (1940). Observations on animals painted with tobacco. *American Journal of Cancer Research, 38*, 41–49.

Summers, K., and Crespi, B. (2008). The androgen receptor and prostate cancer: A role for sexual selection and sexual conflict? *Medical Hypotheses, 70*(2), 435–443.

Sun, T., Plutynski, A., Ward, S., and Rubin, J. B. (2015). An integrative view on sex differences in brain tumors. *Cellular and Molecular Life Sciences, 72*(17), 3323–3342.

Suppes, P. (1970). *A probabilistic theory of causality.* Amsterdam: North-Holland Publishing Company.

Surgeon General's Report. (1964). Smoking and health: Report of the Advisory Committee to the Surgeon General of the Public Health Service. Surgeon General's Advisory Committee on Smoking and Health United States. Public Health Service.

Office of the Surgeon General. *U.S. Department of Health, Education and Welfare, Public Health Service Publication* No. 1103.

Teller, Paul. (2001). Twilight of the perfect model. *Erkenntnis, 55,* 393–415.

Thagard, P. (1999). *How scientists explain disease.* Princeton, NJ: Princeton University Press.

Thiery, J. P. (2002). Epithelial–mesenchymal transitions in tumour progression. *Nature Reviews Cancer, 2*(6), 442.

Tomasetti, C., Marchionni, L., Nowak, M. A., Parmigiani, G., and Vogelstein, B. (2015). Only three driver gene mutations are required for the development of lung and colorectal cancers. *Proceedings of the National Academy of Sciences, 112*(1), 118–123.

Tomasetti, C., and Vogelstein, B. (2015). Variation in cancer risk among tissues can be explained by the number of stem cell divisions. *Science, 347*(6217), 78–81.

Turski, M. L., Vidwans, S. J., Janku, F., Garrido-Laguna, I., et al. (2016). Genomically driven tumors and actionability across histologies: BRAF-mutant cancers as a paradigm. *Molecular cancer therapeutics, 15*(4), 533.

USPSTF. (2009). Nelson, H. D., Tyne, K., Naik, A., Bougatsos, C., et al. (2009). Screening for breast cancer: An update for the US Preventive Services Task Force. *Annals of Internal Medicine, 151*(10), 727–737.

Vandenbroucke, J. P. (1988). Is "the causes of cancer" a miasma theory for the end of the twentieth century? *International Journal of Epidemiology, 17*(4), 708–709.

Van Fraassen, B. C. (1980). *The scientific image.* Oxford: Oxford University Press.

Venkitaraman, A. R. (2001). Functions of BRCA1 and BRCA2 in the biological response to DNA damage. *Journal of Cell Science, 114*(20), 3591–3598.

Virnig, B. A., Tuttle, T. M., Shamliyan, T., and Kane, R. L. (2010). Ductal carcinoma in situ of the breast: A systematic review of incidence, treatment, and outcomes. *JNCI: Journal of the Cancer Institute, 102*(3), 170–178.

Vogelstein, B., and Kinzler, K. (2004). Cancer genes and the pathways they control. *Nature Medicine, 10*(8), 789–799.

Wachbroit, R. (1994). Normality as a biological concept. *Philosophy of Science, 61*(4), 579–591.

Wagner, A. (2013). *Robustness and evolvability in living systems.* Princeton, NJ: Princeton University Press.

Wakefield, J. C. (1992). The concept of mental disorder. *American Psychologist, 47,* 373–388.

Wang, Q., Fan, S., Eastman, A., Worland, P. J., et al. (1996). UCN-01: A potent abrogator of G2 checkpoint function in cancer cells with disrupted p53. *JNCI: Journal of the National Cancer Institute, 88*(14), 956–965.

Wang, F., Hansen, R. K., Radisky, D., Yoneda, T., et al. (2002). Phenotypic reversion or death of cancer cells by altering signaling pathways in three-dimensional contexts. *JNCI: Journal of the National Cancer Institute, 94*(19), 1494–1503.

Warburg, O. (1956). On the origin of cancer. *Science, 123*(3191), 309–314.

Waters, K. (2007a). Molecular genetics. In *Stanford encyclopedia of philosophy* URL: http://plato.stanford.edu/entries/molecular-genetics/ (accessed July 1, 2016)

———. (2007b). Causes that make a difference. *Journal of Philosophy*, *104*(11), 551–579.

Watson, J. D., and Crick, F. H. (1953). Molecular structure of nucleic acids. *Nature*, *171*(4356), 737–738.

Weaver, V. M., Petersen, O. W, Wang, C. A., Larabell, P. B., et al. (1997). Reversion of the malignant phenotype of human breast cells in three-dimensional culture and in vivo by integrin blocking antibodies. *Journal of Cell Biology*, *137*, 231–245.

Weaver, V. M., and Gilbert, P. (2004). Watch thy neighbor: Cancer is a communal affair. *Journal of Cell Science*, *117*, 1287–1290.

Weber, M. (2009). How probabilistic causation can account for the use of mechanistic evidence. *International Studies in the Philosophy of Science*, *23*(3), 277–295.

Weed, D. L. (1986). On the logic of causal inference. *American Journal of Epidemiology*, *123*(6), 965–979.

Weiderpass, E., Adami, H. O., Baron, J. A., Wicklund-Glynn, A., et al. (2000). Organochlorines and endometrial cancer risk. *Cancer Epidemiology, Biomarkers & Prevention*, *9*(5), 487–493.

Weinberg, R. (2014). *The biology of cancer* (2nd Ed.). New York: Garland Science.

Weisberg, M. (2009). The structure of tradeoffs in model building. *Synthese*, *170*(1), 169–190.

Welch, G., and Black, W. (2010). Overdiagnosis in cancer. *JNCI: Journal of the National Cancer Institute*, *102*, 605–613.

Welch, H. G., Schwartz, L., and Woloshin, S. (2011). *Overdiagnosed: Making people sick in the pursuit of health*. Boston: Beacon Press.

Welsh, J. S. (2011). Contagious cancer. *Oncologist*, *16*, 1–4.

Whewell, W. (1837). *The philosophy of the inductive sciences, founded upon their history* (2 Vols.). London: John W. Parker.

———. (1858). *The history of scientific ideas* (2 Vols.). London: John W. Parker.

———. (1860). *On the philosophy of discovery*. London: Chapters Historical and Critical.

Williams, G. C., and Neese, R. M. (1991). The dawn of Darwinian medicine. *Quarterly Review of Biology*, *66*, 1–22.

Williams, N. (2011). Arthritis and nature's joints. In J. K. Campbell, M. O'Rourke, and Slater, M. (Eds.), *Carving nature at its joints: Natural kinds in metaphysics and in science* (199–231). Topics in Contemporary Philosophy (Vol. 8). Cambridge, MA: Bradford Books.

Williams, P., Winzer, K., Chan, W., and Cámara, M. (2007). Look who's talking: Communication and quorum sensing in the bacterial world. *Philosophical Transactions of the Royal Society of London. Series B*, *362*(1483), 1119–1134.

Wilson, D. S. (1975). A theory of group selection. *Proceedings of the National Academy of Sciences, 72*(1), 143–146.

Wilson, R. A. (1999). *Species: New interdisciplinary essays*. MIT Press. A Bradford Book.

Wilson, R. A., Barker, M. J., and Brigandt, I. (2007). When traditional essentialism fails: Biological natural kinds. *Philosophical Topics, 35*(1), 189–215.

Wimsatt, W. C. (1972). Complexity and organization. In *PSA: Proceedings of the Biennial Meeting of the Philosophy of Science Association* (Vol. 20, 67–86). Dordrecht: D. Reidel.

———. (1987). False models as means to truer theories. In M. H. Nitecki and A. Hoffman (Eds.), *Neutral models in biology* (23–55). Oxford: Oxford University Press.

———. (2007). *Re-Engineering philosophy for limited beings: Piecewise approximations to reality*. Cambridge, MA: Harvard University Press.

Wimsatt, W. C., and Schank, J. C. (2004). Generative entrenchment, modularity, and evolvability: When genic selection meets the whole organism. In G. Schlosser and G. P. Wagner (Eds.), *Modularity in development and evolution*. Chicago: University of Chicago Press.

Woodger, J. H. ([1929] 2001). *Biological principles: A critical study* (Vol. 6). London: Routledge and Kegan Paul.

Woodward, J. (1989). The causal mechanical model of explanation. *Minnesota Studies in the Philosophy of Science, 13*, 359–383.

———. (2003). *Making things happen: A theory of causal explanation*. Oxford: Oxford University Press.

———. (2010). Causation in biology: Stability, specificity, and the choice of levels of explanation. *Biology and Philosophy, 25*, 287–318.

———. (2013). Mechanistic explanation: Its scope and limits. *Proceedings of the Aristotelian Society: Supplementary, 87*(1), 39–65.

———. (2014a). Scientific explanation. In E. N. Zalta (Ed.), *The Stanford encyclopedia of philosophy* (2017 Ed.). https://plato.stanford.edu/entries/scientific-explanation/

———. (2014b). A functional account of causation; or, a defense of the legitimacy of causal thinking by reference to the only standard that matters—usefulness (as opposed to metaphysics or agreement with intuitive judgment). *Philosophy of Science, 81*(5), 691–713.

———. (2016). The problem of variable choice. *Synthese, 193*(4), 1047–1072.

———. (2020). Sketch of some themes for a pragmatic philosophy of science.

Worrall, J. (2002). What evidence in evidence based medicine? *Philosophy of Science, 69*, 316–330.

———. (2007). Why there's no cause to randomize. *British Journal for the Philosophy of Science, 58*(3), 451–488.

———. (2011). Causality in medicine: Getting back to the hill top. *Preventive Medicine, 53*(4), 235–238.

Worrall, J., and Worrall, J. (2001). Defining disease: Much ado about nothing? *Analecta Husserliana, 72*, 33–55.

Wright, L. (1973). Functions. *The Philosophical Review, 82*(2), 139–168.

Wynder, E. (1997). Smoking as a cause of lung cancer: Some reflections. *American Journal of Epidemiology, 46*(9), 687–694.

Wydner, E. L., and Graham, E. A. (1950). Tobacco smoking as a possible etiologic factor in bronchiogenic carcinoma: A study of 684 proved cases. *JAMA, 143*(4), 329–336.

Xu, R., Boudreau, A., and Bissell, M. J. (2009). Tissue architecture and function: Dynamic reciprocity via extra-and intra-cellular matrices. *Cancer and Metastasis Reviews, 28*(1–2), 167–176.

Xu, X., Hou, Y., Yin, X., Bao, L., et al. (2012). Single-cell exome sequencing reveals single-nucleotide mutation characteristics of a kidney tumor. *Cell, 148*(5), 886–895.

Yachida, S., Jones, S., Bozic, I., Antal, T., et al. (2010). Distant metastasis occurs late during the genetic evolution of pancreatic cancer. *Nature, 467*(7319), 1114–1117.

Yarden, Y., and Pines, G. (2012). The ERBB network: At last, cancer therapy meets systems biology. *Nature Reviews: Cancer, 12*(8), 553.

AUTHOR INDEX

SUBJECT INDEX

Printed in the USA
CPSIA information can be obtained
at www.ICGtesting.com
CBHW071558151124
17467CB00010B/813